Lecture Notes in Mathematics

Edited by A. Dold and B. Eckmann

507

Michael Reed

Abstract Non Linear Wave Equations

Springer-Verlag
Berlin · Heidelberg · New York 1976

Author
Michael C. Reed
Department of Mathematics
Duke University
Durham, North Carolina 27706
USA

Library of Congress Cataloging in Publication Data

Reed, Michael.
Abstract non-linear wave equations.

(Lecture notes in mathematics ; 507)
Based on lectures delivered at the Zentrum für interdisziplinäre Forschung in 1975.
Bibliography: p.
1. Wave equation. I. Title. II. Series: Lecture notes in mathematics (Berlin) ; 507.
QA3.L28 no. 507 [QC174.26.W3] 510'.8s [530.1'24]
76-2551

AMS Subject Classifications (1970): 35L60, 47H15

ISBN 3-540-07617-4 Springer-Verlag Berlin · Heidelberg · New York
ISBN 0-387-07617-4 Springer-Verlag New York · Heidelberg · Berlin

Printed in Germany

Printing and binding: Beltz, Offsetdruck, Hemsbach/Bergstr.

Preface

These notes cover a set of eighteen lectures delivered at the Zentrum für interdisziplinäre Forschung of the University of Bielefeld in the summer of 1975 as part of the year long project "Mathematical Problems of Quantum Dynamics". It is a pleasure to thank the Zentrum for the opportunity to give these lectures and the Physics faculty of the University of Bielefeld for their warmth and hospitality. Three people deserve special thanks: L. Streit, for extending the invitation, C. Pfister, for help in the preparation of the manuscript, and M. Kämper for her excellent typing.

Mike Reed

Bielefeld, August, 1975

Table of Contents

Introduction

During the past year there has been a great deal of interest, both in applied physics and in quantum field theory, in non-linear wave equations. Like all non-linear problems, these equations must to some extent be dealt with individually because each equation has its own special properties. Thus, in the literature these equations are often treated separately; the proofs of existence and properties of solutions often seem to depend on special properties of the particular equation studied. In fact, these equations have in common certain basic problems in abstract non-linear functional analysis. Using just the standard tools of linear functional analysis and the contraction mapping principle, one can go quite far on the abstract level, thus providing a unified approach to these non-linear equations. Furthermore, the abstract approach makes it clear which properties of the solutions are general and which depend on special properties of the equations themselves. The abstract approach, which originated in the work of Segal [31], has been neglected partially because one can always push further in a particular case using special properties. Nevertheless, the abstract methods and ideas form the core of much of the recent work, although this is somewhat obscured in the literature. Furthermore, there are many beautiful and important unsolved problems both on the abstract and the particular level. For all of these reasons, it seems an appropriate time to pull together what is known about the abstract theory and how it is applied.

To see how the abstract questions arise naturally from an example, consider the non-linear Klein-Gordon equation:

$$(1) \qquad u_{tt}(x,t) - \Delta u(x,t) + m^2 u(x,t) = -g|u(x,t)|^{p-1}u(x,t) \qquad x \in R^n$$

$$u(x,o) = f(x)$$

$$u_t(x,o) = g(x)$$

Basically, one would like information about local existence of solutions, global existence, smoothness, finite propagation speed, continuous dependence on the initial data and eventually we want a scattering theory. As we will see, the results and the techniques depend critically on the size of p and n and the sign of g. To treat (1) in a general setting we reformulate it as a first order system as follows:

Let $v(x,t) = u_t(x,t)$, then

(2) $$v_t - \Delta u + m^2 u = - g|u|^{p-1}u$$

$$u_t = v$$

$$u(x,o) = f(x)$$

$$v(x,o) = g(x)$$

Now, for each t, define

(3) $\phi(t) = \langle u(x,t), v(x,t)\rangle \qquad J(\phi(t)) = \langle o, - g|u|^{p-1}u\rangle$

Then, we can rewrite (2) as

(4) $$\phi'(t) - \begin{pmatrix} o & I \\ \Delta - m^2 & o \end{pmatrix}\phi(t) = J(\phi(t))$$

$$\phi(o) = \langle f(x), g(x)\rangle$$

The operator $-\Delta + m^2$ is a positive self-adjoint operator on $L^2(R^n)$ with domain given by those $f \in L^2(R^n)$ so that

$$(k^2 + m^2)\hat{f}(k) \in L^2(R^n)$$

where ^ always denotes the Fourier transform. We denote by B the unique positive self-adjoint square root of $-\Delta + m^2$ given by the functional calculus. Since B is strictly positive and closed, D(B) is a Hilbert space under the inner product $(Bu, Bu)_{L^2}$. We can thus define the Hilbert space $\mathcal{H}_B = D(B) \oplus L^2(R^n)$ with the inner product

$$(\langle u, v\rangle, \langle u, v\rangle)_B = (Bu, Bu)_{L^2} + (v, v)_{L^2}$$

Now let

(5) $$A = i\begin{pmatrix} o & I \\ -B^2 & o \end{pmatrix}$$

Then, one can check that A is a symmetric operator on $\mathcal{H}_B$ with domain $D(A) = D(B^2) \oplus D(B)$ and A is closed since B and B^2 are closed.

Now define for each t,

$$W(t) = \begin{pmatrix} \cos(tB) & B^{-1}\sin(tB) \\ -B\sin(tB) & \cos(tB) \end{pmatrix}$$

where each of the entries is defined by the functional calculus for B on $L^2(R^n)$. One can now compute directly that $W(t)$ is a strongly continuous one-parameter group and that for $\psi \in D(A)$ the strong derivative of $W(t)\psi$ exists and equals $-iA\psi$. Since $W(t)$ takes $D(A)$ into itself, a corollary of Stone's theorem (see [26],p.269) says that A is self-adjoint in D(A) and generates $W(t)$. We have thus reformulated our original problem (1) as follows. We must find an $\mathcal{H}_B$ - valued function $\phi(\cdot)$ on R which satisfies:

$$\frac{d\phi(t)}{dt} = -iA\phi(t) + J(\phi(t))$$

$$\phi(o) = \phi_o \equiv \langle f,g \rangle$$

where the self-adjoint operator A is given by (5) and J is given by (3).

This shows that (1) is really a special case of a very general class of Hilbert space problems. Namely, given a self-adjoint operator on a Hilbert space $\mathcal{H}$, a vector ϕ_o in $\mathcal{H}$, and a non-linear mapping J of $\mathcal{H}$ into itself, when can one find an $\mathcal{H}$-valued function $\phi(\cdot)$ on R which solves the initial-value problem:

$$\phi'(t) = -iA\phi(t) + J(\phi(t)) \tag{6}$$

$$\phi(o) = \phi_o$$

It is this abstract problem which is the main subject of these lectures, but the main motivation for studying the abstract problem is to prove classical properties (existence, smoothness, etc.) of non-linear wave equations. So, we will always return to the equation (1) and to other non-linear equations like the sine-Gordon equation, the coupled Dirac-Klein-Gordon equations, the non-linear Schrödinger equation. Our treatment of applications is uneven because there is no attempt to be complete - the applications are used to show the different ways the abstract theory can be applied. Thus we sometimes provide most of the details and sometimes just sketch the important features or say how one application differs from other. Furthermore, there are many beautiful applications which we don't even mention but which can be found in the papers listed in the bibliography.

The material of the lectures falls naturally into two parts. In Chapter 1 we treat the existence theory and properties of solutions of (6). Here the abstract theory is quite complete and the method of application well understood. Nevertheless, there are many unsolved problems, either in the details of specific applications or in extending the theory to cover new applications. These problems, some of which are very difficult will become clear as we proceed.

In Chapter 2 we develop various aspects of a scattering theory for (6) which roughly means comparing solutions of (6) to solutions of the "free" equation

$$\phi'(t) = - iA\phi(t)$$
$$\phi(o) = \tilde{\phi}_o$$

for large positive and negative times. Here the abstract theory is less satisfactory in that many more hypotheses on A and J are required, the applications are less well understood, and more abstract results are required in order to have a complete theory. If I can arouse your interest in these many unsolved problems then I will have achieved my main purpose.

Chapter 1 Existence and Properties of Solutions

1. Local and Global Existence

In this section we prove a local existence theorem for (6) and then give various applications. The basic idea is the same as the proof of existence of solutions of ordinary differential equations. We reformulate (6) as an integral equation problem:

$$\phi(t) = e^{-iAt}\phi_0 + \int_0^t e^{-iA(t-s)}J(\phi(s))ds \tag{7}$$

and then solve (7) by the contraction mapping principle.

Theorem 1 (local existence). Let A be a self-adjoint operator on a Hilbert space $\mathcal{H}$ and J a mapping from D(A) to D(A) which satisfies:

(H_0) $\|J(\phi)\| \leq C(\|\phi\|)\|\phi\|$

(H_1) $\|AJ(\phi)\| \leq C(\|\phi\|, \|A\phi\|)\|A\phi\|$

(H_0^L) $\|J(\phi) - J(\psi)\| \leq C(\|\phi\|, \|\psi\|)\|\phi - \psi\|$

(H_1^L) $\|A(J(\phi) - J(\psi))\| \leq C(\|\phi\|, \|A\phi\|, \|\psi\|, \|A\psi\|)\ \|A\phi - A\psi\|$

for all $\phi,\psi\in D(A)$ where each constant C is a monotone increasing (everywhere finite) function of the norms indicated. Then, for each $\phi_0 \in D(A)$ there is a $T > 0$ so that (6) has a unique continuously differentiable solution for $t\in[0,T)$. For each set of the form

$$\{\phi \,|\, \|\phi\| \leq a,\ \|A\phi\| \leq b\},$$

T can be chosen uniformly for all ϕ_0 in the set.

Proof. Let X(T) be the set of D(A) -valued functions on $[0,T)$ for which $\phi(t)$ and $A\phi(t)$ are continuous and

$$\|\phi(\cdot)\|_T \equiv \sup_{t\in[0,T)} \|\phi(t)\| + \sup_{t\in[0,T)} \|A\phi(t)\| < \infty$$

Since A is a closed operator, X(T) with the norm $||\phi(\ )||_T$ is a Banach space. Choose some fixed $\alpha > 0$. Let $\phi_o \in D(A)$ be given and let $X(T,\alpha,\phi_o)$ consist of those $\phi(\ )$ in X(T) with $\phi(0) = \phi_o$ and $||\phi(\ ) - e^{iAt}\phi_o||_T \le \alpha$. We will show that the map

$$(S\phi)(t) = e^{-iAt}\phi_o + \int_0^t e^{-iA(t-s)} J(\phi(s))ds \tag{8}$$

is a contraction on $X(T,\alpha,\phi_o)$ if T is small enough. We denote by C_α any of the constants in the hypotheses with arguments $||\phi_o|| + \alpha$ and $||A\phi_o|| + \alpha$. Suppose that $\phi(\cdot)\in X(T,\alpha,\phi_o)$; then

$$||e^{-iA(t-(s+h))}J(\phi(s+h)) - e^{-iA(t-s)}J(\phi(s))||$$

$$\le ||J(\phi(s+h)) - J(\phi(s))|| + ||(e^{-iAh} - I)\ J(\phi(s))||$$

$$\le C_\alpha||\phi(s+h) - \phi(s)|| + ||(e^{-iAh} - I)\ J(\phi(s))||$$

so $e^{-iA(t-s)}J(\phi(s))$ is a continuous $\mathcal{H}$-valued function of s. A similar proof shows that $Ae^{-iA(t-s)}J(\phi(s))$ is also continuous. Thus, the right-hand side of (8) can be defined using the Riemann integral, and if

$$\eta_n(t) \equiv \sum_{m=1}^{n} \frac{1}{n}e^{-i(t-(m/n)t)A}\ J(\phi(\tfrac{m}{n}t))$$

and

$$\eta(t) \equiv \int_0^t e^{-i(t-s)A}\ J(\phi(s))ds$$

then $\eta_n(t) \to \eta(t)$ as $n \to \infty$. Now, by the hypotheses on J, each $\eta_n(t)\in D(A)$, so

$$A\eta_n(t) = \sum_{m=1}^{n} \frac{1}{n}e^{-i(t-(m/n)t)A}\ AJ(\phi(\tfrac{m}{n}t))$$

$$\to \int_0^t e^{-i(t-s)A}\ AJ(\phi(s))ds$$

Therefore, $\eta(t)\epsilon D(A)$ and

$$(9) \qquad A\int_0^t e^{-i(t-s)A} J(\phi(s))ds = \int_0^t e^{-i(t-s)A} AJ(\phi(s))ds$$

Further,

$$||A\eta(t+h) - A\eta(t)|| \leq \left|\left|\int_t^{t+h} e^{-iA(t-s)} e^{-iAh} AJ(\phi(s))ds\right|\right|$$

$$+ \left|\left|\int_0^t (e^{-ihA} - I) e^{-iA(t-s)} AJ(\phi(s))ds\right|\right|$$

$$\leq hC_\alpha ||\phi||_T + \int_0^t ||(e^{-ihA} - I)AJ(\phi(s))||ds$$

The integrand in the second term converges to zero as $h\to 0$ for each s and by the hypotheses on J, the integrand is uniformly bounded. Thus, by the dominated convergence theorem the right-hand side converges to zero as $h\to 0$, so $A\eta(t)$ is continuous and similarly, $\eta(t)$ is continuous. Further, exactly the same kind of estimates as above show that for any $\phi(\cdot)$ and $\psi(\cdot)\epsilon X(T,\alpha,\phi_0)$, we have

$$||(S\phi)(t) - e^{-iAt}\phi_0|| \leq C_\alpha T \sup_{t\in[0,T)} ||\phi(t)||$$

$$||A(S\phi)(t) - Ae^{-iAt}\phi_0|| \leq C_\alpha T \sup_{t\in[0,T)} ||A\phi(t)||$$

$$||(S\phi)(t) - (S\psi)(t)|| \leq C_\alpha T \sup_{t\in[0,T)} ||\phi(t) - \psi(t)||$$

$$||A\big((S\phi)(t) - (S\psi)(t)\big)|| \leq C_\alpha T \sup_{t\in[0,T)} ||A\phi(t) - A\psi(t)||$$

Thus, for T small enough, S is a contraction on $X(T,\alpha,\phi_0)$ so, by the contraction mapping principle (see [26], p, 151), S has a unique fixed point $\Phi(\cdot)$ in $X(T,\alpha,\phi_0)$ which satisfies (7). Now, suppose that $\tilde{\phi}$ is a continuoulsy differentiable D(A) -valued solution of (6) on the interval $[0,\tilde{T})$ with $\tilde{\phi}(0) = \phi_0$. By the differential equation, $A\tilde{\phi}(t)$ is continuous, so $\tilde{\phi}(t) \in X(T_0,\alpha,\phi_0)$ for t in some interval $[0,T_0)$. Since $\tilde{\phi}$ obeys (7) $\phi(t) = \tilde{\phi}(t)$ for $t < T_0$. Let T_1 be the sup of such T_0. Then, if $T_1 < T$, $\lim_{t\to T_1}\tilde{\phi}(t) = \phi(T_1)$ and $\lim_{t\to T_1}A\tilde{\phi}(t) = A\phi(T_1)$ exist so $\tilde{\phi}(T_1) = \phi(T_1) \in D(A)$ and the same argument as above shows that $\tilde{\phi}(t) = \phi(t)$ for some small interval $T_1 \leq t < T_2 < T$ which contradicts the maximality of T_1. Thus $T_1 \geq T$, so $\tilde{\phi}(t) = \phi(t)$ for $t \in [0,T)$. That is, any strong solution of (6) on $[0,T)$ equals $\phi(t)$.

To prove the strong differentiability of $\phi(t)$, we write

$$\frac{\phi(t+h) - \phi(t)}{h} = \left(\frac{e^{-iAh} - I}{h}\right) e^{-iAt}\phi_0 + \frac{1}{h}\int_t^{t+h} e^{-iA(t-s)}e^{-ihA}J(\phi(s))ds$$

$$+ \int_0^t e^{-iA(t-s)}\left(\frac{e^{-ihA} - I}{h}\right) J(\phi(s))ds \tag{10}$$

Since $\phi_0 \in D(A)$ the first term converges to $-iAe^{-iAt}\phi_0$ as $h \to 0$ and since the integrand of the second term is continuous, it converges to $J(\phi(t))$. The integrand of the third term converges to $e^{-iA(t-s)}(-iAJ(\phi(s)))$ for each s and

$$\left\| \frac{e^{-ihA} - I}{h} J(\phi(s)) \right\| \leq \|AJ(\phi(s))\| \leq C\|A\phi_0\| + \alpha$$

so the integrand is uniformly bounded. Thus, by the dominated convergence theorem, the third term converges as $h \to 0$ to

$$\int_0^t e^{-iA(t-s)} (-iAJ(\phi(s)))ds$$

which by (9) equals

$$-iA\int_0^t e^{-iA(t-s)} J(\phi(s))ds$$

Therefore $\phi(t)$ is strongly differentiable for $t \in [0,T)$ and satisfies (6). ∎

Corollary 1 Let A be a self-adjoint operator on $\mathcal{H}$, and suppose that and non-linear mapping J takes $\mathcal{H}$ into itself and satisfies (H_o^L). Then for each $\phi_o \in \mathcal{H}$, there is a $T > o$ such that (7) has a continuous solution $\phi(t)$ on $[o,T)$. T can be chosen uniformly on balls in $\mathcal{H}$.

Proof This is really a corollary of the proof of Theorem 1. The proof is similar, only much easier since we don't have to differentiate the integral equation. Just define

$$X(T,\alpha,\phi_o) = \{ \psi(t) \mid \sup_{t \in [o,T)} \| \psi(t) - e^{-iAt}\phi_o \| \leq \alpha,\ \psi(o) = \phi_o \}$$

and proceed as before. ∎

We remark that the hypotheses (H_o) and (H_1) follow from hypotheses (H_o^L) and (H_1^L). We state them separately for easy comparison with the hypotheses Theorem 2 below. Notice also that the same proof shows local existence on an interval $(-T,T)$ since e^{-iAt} is a group.

It is well known from freshman calculus that non-linear ordinary differential equations may not have global solutions in time (for example: $\frac{dx}{dt} = x^2$). As in that case, one can prove global existence if one has as apriori estimate which guarantees that the solution remains bounded.

Theorem 2 (global existence) Let A, J, and $\mathcal{H}$ satisfy the hypotheses of Theorem 1 except that (H_1) is replaced by

$$(H_1') \qquad \|AJ(\phi)\| \leq C(\|\phi\|)\|A\phi\|$$

(i.e. C does not depend on $\|A\phi\|$). Suppose that on every finite interval on which a strong solution of (6) exists, $\|\phi(t)\|$ is bounded. Then (6) has a unique global solution for all time.

Proof: By Theorem 1 we know that a solution exists on $[o,T)$ for some $T > o$ small enough. Let $\bar{T}$ be the supremum of the numbers T so that a solution of (6) exists on $[o,T)$ with $\phi(t)$ and $A\phi(t)$ continuous. Suppose that $\bar{T} < \infty$ and $\|A\phi(t)\|$ is bounded on $[o,\bar{T})$. Now in the local existence proof in Theorem 1 the length of the interval of existence depended only on the numbers $\|\phi_o\|$, $\|A\phi_o\|$. Therefore, if we choose a point T_1 close enough to $\bar{T}$ we can construct a solution of (6) on $[T_1, T_2)$ where $T_1 < \bar{T} < T_2$. By local uniqueness this solution extends our previous solution on $[o,\bar{T})$ and thus violates the maximality of $\bar{T}$. Therefore $\bar{T} = \infty$, so the solution exists on $[o,\infty)$ and a similar proof shows that it exists on the interval $(-\infty,o]$.

It remains to show that $||A\phi(t)||$ is apriori bounded on any finite interval where the solution exists. The solution of (6) satisfies (7) on any such interval, so

$$||A\phi(t)|| \leq ||A\phi_0|| + \int_0^t ||e^{-iA(t-s)} AJ(\phi(s))||ds$$

$$\leq ||A\phi_0|| + \int_0^t C(||\phi(s)||)\, ||A\phi(s)||ds$$

by (H_1'). By hypothesis, $||\phi(s)||$ is bounded on any finite interval $[o,T)$ where the solution exists, say by $C_1(T)$. Since $C(\cdot)$ is locally bounded we have

$$||A\phi(t)|| \leq ||A\phi_0|| + C(C_1(T)) \int_0^t ||A\phi(s)||ds$$

for all $t < T$. By iteration this implies that

$$||A\phi(t)|| \leq ||A\phi_0||e^{tC(C_1(T))}$$

which shows that $||A\phi(t)||$ is bounded on $[o,T)$ if T is finite. ∎

The original idea of using this abstract formulation to prove existence for non linear wave equations is due to I.Segal[31]. Theorems 1 and 2 and also the theorems in Section 3 are simplified versions of his ideas (see also Reed[24]and Reed and Simon [26]). This is further discussed in the notes to Section 3. Other hypotheses on J and A guaranteeing local and global existence can be found in Browder [2]. W. von Wahl [40] treats the case where A depends on t.

2. Applications

In this section we show how the existence theorems may be applied. Usually, proving hypotheses (H_i) , (H_i^L) , $i = 0,1$, depends on Sobolev estimates and sometimes on long and rather tedious calculations. In general, the estimates are verified on a dense set of nice vectors for A and then one passes to the full domain of A by a limiting argument. Finally, since we are dealing with vector-valued functions some extra arguments are necessary to use energy inequalities. Although many of these details are of a "technical" nature, it is important to understand how to carry them through.

Therefore, in our first example we will do most of the tedious technical details and prove the requisite Sobolev estimate. After that we will just quote known estimates and say that the technical details are similar to the first example.

(A) $\underline{u_{tt} - \Delta u + m^2 u = - \lambda |u|^{p-1} u , \quad \lambda > 0, \quad m > 0, \quad n = 3, \quad p = 3}$

For convenience we restate the basic definitions:

$$\mathcal{H} = \{ \langle u,v\rangle \mid u,v \in L^2(R^3), \ \|\langle u,v\rangle\|^2 \equiv \|Bu\|_2^2 + \|v\|_2^2 < \infty \}$$

where $B = \sqrt{-\Delta + m^2}$. For $\phi = \langle u,v\rangle \in \mathcal{H}$, we define $J(\phi) = \langle 0, -\lambda|u|^2 u\rangle$ Notice that $\|\cdot\|$ always denotes the norm on $\mathcal{H}$ and $\|\cdot\|_p$ will denote the usual L^p norm on R^3. As pointed out in the introduction,

$$A = i\begin{pmatrix} 0 & I \\ -B^2 & 0 \end{pmatrix}$$

is self-adjoint on

$$D(A) = \{ \langle u,v\rangle \in \mathcal{H} \mid u \in D(B^2), \quad v \in D(B) \}$$

All of the estimates in this example are based on the following simple Sobolev inequality. We denote the Fourier transform of a function f by $\hat{f}$ and various universal constants will be denoted by K.

Lemma 1 Let $u \in C_0^\infty(R^3)$. Then $||u||_6 \le K||Bu||_2$

Proof: Denote $\partial u(x)/\partial x_i$ by $\partial_i u$. Then by the fundamental theorem of calculus,

$$|u(x)|^4 \le 4\int |u^3 \partial_i u| dx_i$$

where the integral is taken over the line where x_j is held fixed for $j \neq i$. Thus,

$$|u(x)|^6 \le K\left(\int |u^3\partial_1 u| dx_1\right)^{1/2} \left(\int |u^3 \partial_2 u| dx_2\right)^{1/2} \left(\int |u^3\partial_3 u| dx_3\right)^{1/2}$$

By integrating both sides (by iterating the integrals) and using the Schwarz inequality, one obtains

$$\int_{R^3} |u|^6 dx \le K\left(\int_{R^3} |u^3 \partial_1 u| dx\right)^{1/2} \left(\int_{R^3} |u^3\partial_2 u| dx\right)^{1/2} \left(\int_{R^3} |u^3 \partial_3 u| dx\right)^{1/2}$$

$$\le K\left(\int_{R^3} |u|^6 dx\right)^{3/4} \left(\int_{R^3} |\partial_1 u|^2 dx\right)^{1/4}$$

$$\cdot \left(\int_{R^3} |\partial_2 u|^2 dx\right)^{1/4} \left(\int_{R^3} |\partial_3 u|^2 dx\right)^{1/4}$$

From this one easily obtains

$$\left(\int_{R^3} |u|^6 dx\right)^{1/6} \le K(||\partial_1 u||_2 + ||\partial_2 u||_2 + ||\partial_3 u||_2)$$

$$= K(||k_1 \hat{u}||_2 + ||k_2 \hat{u}||_2 + ||k_3 \hat{u}||_2)$$

$$\le K||(\textstyle\sum k_i^2 + m^2)^{1/2} \hat{u}||_2$$

$$= K||Bu||_2 \quad \blacksquare$$

To extend this estimate to D(B), we need:

Lemma 2: B is essentially self-adjoint on $C_0^\infty(R^3)$.

Proof: By using the Fourier transform it is clear that both B and B^2 are essentially self-adjoint on the Schwarz space $\mathcal{S}(R^n)$. Now let $u \in \mathcal{S}(R^n)$ be given. Since $C_0^\infty(R^3)$ is dense in $\mathcal{S}(R^n)$ (in the $\mathcal{S}(R^n)$ topology) we can find $u_n \in C_0^\infty(R^3)$ so that $u_n \xrightarrow{L^2} u$ and $B^2 u_n \xrightarrow{L^2} B^2 u$. But,

$$||B(u_n - u)|| \le ||(B^2 + I)(u_n - u)||$$

$$\le ||B^2(u_n - u)|| + ||u_n - u||$$

$$\longrightarrow 0$$

Therefore B restricted to $\mathcal{S}(R^n)$ is in the closure of B restricted to $C_0^\infty(R^n)$. Thus, B is essentially self-adjoint on $C_0^\infty(R^3)$. ∎

Lemma 3 Suppose that $u_1, u_2, u_3 \in D(B)$. Then

$$||u_1 u_2 u_3||_2 \le K||Bu_1||_2||Bu_2||_2||Bu_3||_2$$

Proof Let $u \in D(B)$. Since B is essentially self-adjoint on $C_0(R^3)$, we can find a sequence of $C_0^\infty(R^3)$ functions u_n so that $u_n \xrightarrow{L^2} u$, and $Bu_n \xrightarrow{L^2} Bu$, and by passing to a subsequence if necessary, we may assume u_n converges pointwise to u also. But

$$||u_n^3 - u_m^3||_2 = ||(u_n - u_m)(u_n^2 + u_n u_m + u_m^2)||_2$$

$$\le K||u_n - u_m||_6||(u_n^2 + u_n u_m + u_m^2)||_3$$

$$\le K||u_n - u_m||_6(||u_n||_6^2 + ||u_n||_6||u_m||_6 + ||u_m||_6^2)$$

$$\le K||Bu_n - Bu_m||_2(||Bu_n||_2^2 + ||Bu_n||_2||Bu_m||_2 + ||Bu_m||_2^2)$$

so $\{u_n^3\}$ is Cauchy in L^2 and since it converges pointwise to u^3 we have $u^3 \in L^2$. Taking the limit in the inequality we obtain

$$||u||_6^3 = ||u^3||_2 \leq K||Bu||_2^3$$

The statement of the lemma now follows by applying Hölder's inequality twice. ∎

Lemma 4 For all $\phi_1, \phi_2 \in \mathcal{H}$, J satisfies

$$||J(\phi_1)|| \leq K||\phi_1||^3$$

$$||J(\phi_1) - J(\phi_2)|| \leq C(||\phi_1||, ||\phi_2||)||\phi_1 - \phi_2||$$

Proof: Let $\phi_i = \langle u_i, v_i \rangle$. Then, by Lemma 3,

$$||J(\phi_1)|| = ||\lambda u_1^2 \bar{u}_1||_2 \leq K||Bu_1||_2^3 \leq K||\phi_1||^3$$

and (by the calculation in Lemma 3)

$$||J(\phi_1) - J(\phi_2)|| = ||\lambda(u_1^2\bar{u}_1 - u_2^2\bar{u}_2)||_2$$

$$\leq K||B(u_1 - u_2)||_2(||Bu_1||_2^2 + ||Bu_1||_2||Bu_2||_2 + ||Bu_2||_2^2)$$

$$\leq K||\phi_1 - \phi_2||(||\phi_1||^2 + ||\phi_1|| \, ||\phi_2|| + ||\phi_2||^2)$$

which proves the lemma. ∎

Lemma 5 Let $\phi_1, \phi_2 \in D(A)$, then

$$||AJ(\phi_1)|| \leq K||\phi_1||^2||A\phi_1|| \tag{11}$$

$$||A(J(\phi_1) - J(\phi_2))|| \leq C(||\phi_1||, ||\phi_2||, ||A\phi_1||, ||A\phi_2||)||A\phi_1 - A\phi_2||$$

Proof Let $\phi_i = \langle u_i, v_i \rangle$ where $u_i \in D(B^2)$, $v_i \in D(B)$. We compute

$$||B\partial_i u||_2^2 = ||(\sum k_i^2 + m^2)^{1/2} k_i \hat{u}||_2^2 \leq ||(\sum k_i^2 + m^2)\hat{u}||_2^2 = ||B^2 u||_2^2$$

so, by Lemma 3,

$$||\partial_i(u^2\bar{u})||_2 = ||2uu\partial_i\bar{u} + u^2\partial_i\bar{u}||_2 \leq K||Bu||_2^2||B\partial_i u||_2 \leq K||Bu||_2^2||B^2 u||_2$$

Thus,

$$||AJ(\phi_1)||^2 = \lambda^2||Bu_1^2\bar{u}_1||_2^2 = \lambda^2\sum_{i=1}^{3}||\partial_i(u_1^2\bar{u}_1)||_2^2 + \lambda^2 m^2||u_1^2\bar{u}_1||_2^2$$

$$\leq K(||Bu_1||_2^4||B^2u_1||_2^2 + m^2||Bu_1||_6^2)$$

$$\leq K||Bu_1||_2^4||B^2u_1||_2^2$$

$$\leq K||\phi_1||^4||A\phi_1||^2$$

which proves the first inequality. To prove the second, we compute:

$$||\partial_i(u_1^2\bar{u}_1 - u_2^2\bar{u}_2)||_2^2 \leq ||u_1^2\partial_i(\bar{u}_1 - \bar{u}_2)||_2^2 + ||(u_1^2 - u_2^2)\partial_i(\bar{u}_2)||_2^2$$

$$+ ||2(|u_1|^2 - |u_2|^2)\ \partial_i u_1||_2^2 + ||2|u_2|^2\partial_i(u_1 - u_2)||_2^2$$

$$\leq K(||Bu_1||_2^4||B^2(u_1 - u_2)||_2^2$$

$$+||B^2u_2||_2^2||B(u_1 - u_2)||_2^2||B^2(u_1 - u_2)||_2^2)$$

$$\leq K(||\phi_1||^4||A(\phi_1 - \phi_2)||^2$$

$$+ ||A\phi_2||^2(||\phi_1|| + ||\phi_2||)||A(\phi_1 - \phi_2)||^2$$

Therefore,

$$||A(J(\phi_1) - J(\phi_2))||^2 = \lambda^2||B(u_1^2\bar{u}_1 - u_2^2\bar{u}_2)||_2^2$$

$$= \lambda^2 \sum_{i=1}^{3} ||\partial_i (u_1^2\bar{u}_1 - \bar{u}_2 u_2^2)||_2^2$$

$$+ \lambda^2 m^2 ||u_1^2\bar{u}_1 - u_2^2\bar{u}_2||_2^2$$

$$\leq C(||\phi_1||, ||\phi_2||, ||A\phi_2||)||A(\phi_1 - \phi_2)||^2$$

$$+ C(||\phi_1||, ||\phi_2||)||A(\phi_1 - \phi_2)||^2$$

which proves the lemma. We have several times used the inequality

$$||Bu||_2 \leq K||B^2 u||_2 \quad \blacksquare$$

The last two lemmas show that A, J, and $\mathcal{H}$ satisfy the hypotheses of Theorem 1 so local solutions exist. Notice that in Lemma 5 (formula (11)) we have actually shown the stronger hypotheses (H_1') of Theorem 2. Thus, to prove global existence we need just show that $||\phi(t)||$ is uniformly bounded. Let $\phi(t) = \langle u(t), v(t)\rangle$ be a local solution of (6) with initial data $\phi_o = \langle f,g\rangle \in D(A)$. Then $u(x,t) \in D(B^2)$ and $u_t(x,t) = v(x,t) \in D(B)$ as functions of x for each fixed t. We now define the "energy"

$$E(t) \equiv \frac{1}{2} \int \{|Bu(x,t)|^2 + |u_t(x,t)|^2 + \frac{\lambda}{2}|u(x,t)|^4\} d^3x$$

Since $u \in L^2(R^3) \cap L^6(R^3)$ we have $u \in L^4(R^3)$ and since Bu and u_t are also in L^2 the integral on the right side makes sense. Now, from Theorem 1 we know that $\phi(t)$ is strongly differentiable as an $\mathcal{H}$-valued function which means that

$$||B\left(\frac{u(t+h) - u(t)}{h} - u_t(t)\right)||_2 \longrightarrow 0 \qquad (12)$$

$$\left\| \left(\frac{u_t(t+h) - u_t(t)}{h}\right) - u_{tt}(t) \right\|_2 \longrightarrow 0$$

as $h \longrightarrow 0$. Thus the first two terms in $E(t)$ are differentiable. To see that the third term is differentiable note that

$$\left\| u(t)\left(\frac{u(t+h) - u(t)}{h} - u_t(t)\right)\right\|^2 \leq \|u\|_2^{1/2}\|Bu\|_2^{1/2}\left\|B\left(\frac{u(t+h)-u(t)}{h} - u_t\right)\right\|_2$$

By (12), the right hand side goes to zero as $h \to 0$ and from this it easily follows that $u(t)^2$ is strongly differentiable. Therefore $(u(t)^2, u(t)^2)_2$ is differentiable so the third term in $E(t)$ is differentiable and

$$E'(t) = \frac{1}{2}(Bu_t, Bu) + \frac{1}{2}(u_{tt}, u_t) + \frac{\lambda}{2}(uu_t, u^2)$$

$$+ \frac{1}{2}(Bu, Bu_t) + \frac{1}{2}(u_t, u_{tt}) + \frac{\lambda}{2}(u^2, uu_t)$$

$$= \frac{1}{2}(u_t, B^2u + u_{tt} + \lambda|u|^2u) + \frac{1}{2}(B^2u + u_{tt} + \lambda|u|^2u, u_t)$$

$$= 0$$

since u satisfies $B^2u + u_{tt} + \lambda|u|^2u = 0$.

Now we can show that $\|\phi(t)\|$ is bounded on any finite interval where the solution exists. Since $\lambda > 0$,

$$\frac{1}{2}\|\phi(t)\|^2 \leq \frac{1}{2}\|\phi(t)\|^2 + \frac{\lambda}{4}\int_{R^3}|u(x,t)|^4 dx$$

$$= E(t) = E(0)$$

Thus we have verified the hypotheses of Theorem 2 and so we have the following theorem.

Theorem 3 Suppose $\lambda > 0$, $m > 0$ and $f \in D(-\Delta + m^2)$, $g \in D((-\Delta + m^2)^{1/2})$ in $L^2(R^3)$. Then there exists a unique function $u(x,t)$, $t \in R, x \in R^3$, so that $t \longmapsto u(\cdot,t)$ is a twice continuously strongly differentiable $L^2(R^3)$-valued function of t, $u(\cdot,t) \in D(-\Delta + m^2)$ for all t, $u(x,o) = f(x)$, $u_t(x,o) = g(x)$ and

$$u_{tt} - \Delta u + m^2 u = -\lambda |u|^2 u \tag{13}$$

We make several remarks. Notice that if f and g are real valued then both $u(x,t)$ and $\overline{u(x,t)}$ satisfy (12) so by uniqueness $u = \bar{u}$. Thus u is real and satisfies

$$u_{tt} - \Delta u + m^2 u = -\lambda u^3$$

Secondly, the global existence result depended strongly on the fact that $\lambda \geq o$ so that the term in the conserved energy due to the non-linearity is positive. We will see later (part E) in an analogous example that if this condition fails then global existence doesn't hold. If the non-linearity is mild then one can sometimes prove global existence even though the contribution to the energy of the non-linear term is not positive (see parts D and F below)

Notice that the global existence result also depended on the assumption $\phi_o \in D(A)$. For a general $\phi_o \in \mathcal{H}$, we can solve the integral equation (7) locally by Corollary 1 since the hypothesis (H_o^L) is satisfied. But we cannot show that the solution of (7) is global. The reason is that although $E(t)$ makes sense, we cannot differentiate it since $\phi(t)$ will not (in general) be strongly differentiable. Thus we need some kind of continuity argument to show that $E(t)$ remains bounded since we know it is true for a dense set. We return to this question in Section 4 when we discuss continuous dependence on the data.

Finally, it is important to realize that one has both gained and lost by treating the Klein-Gordon equation in the abstract framework. What we have gained is that the proof was quite easy. Given the general abstract results (Theorems 1 and 2) and the (non-trivial) Sobolev inequality, the rest of the proof was just technical functional analysis details. On the other hand the result of Theorem 3 is not entirely satisfactory. We would like to know if we can solve (13) in a completely classical sense. For example if we start with C^∞ data, will the solution remain C^∞ ? The answer is yes but more work (and estimates) are required (see section 3).

The first proof of global existence of (13) was due to Jürgens [15]. It was his work that stimulated Segal to develope the abstract approach in [31]. The technical details which we have presented are taken from Reed and Simon [27].

B. The case m = O

The case $m = o$ can be handled by the following device. We choose some $m_o > o$ and write the equation as

$$u_{tt} - \Delta u + m_o^2 u = - \lambda u^3 + m_o^2 u \tag{14}$$

$\mathcal{H}$ and A are defined as in part A but J is now defined as

$$J(\langle u, v\rangle) = \langle o, - \lambda |u|^2 u + m_o^2 u\rangle$$

The addition of the linear term in J does not affect any of the estimates on J (except by changing constants) so local existence in guaranteed as before by Theorem 1. Of course by rewriting the equation as (14) we have not changed the conserved energy which is:

$$E(t) \equiv \frac{1}{2} \int_{R^3} \{ |\nabla u|^2 + |u_t|^2 + \frac{\lambda}{2} |u|^4 \} dx$$

As in part A, when $\phi_o \in D(A)$, $E(t)$ is differentiable and $E'(t) = o$. Thus, $E(t) = E(o)$ and from this it follows that

$$\int_{R^3} \{ |\nabla u|^2 + |u_t|^2 \} \, dx \leq 2E(o) \tag{15}$$

However, in order to conclude that $||\phi(t)||^2 = ||Bu(t)||_2^2 + ||u_t(t)||_2^2$ remains bounded on finite intervals we also need that $||u(t)||_2$ remains bounded. By the fundamental theorem of calculus,

$$u(t) = u(o) + \int_o^t u_s(s)ds$$

since u(t) is a strongly continuously differentiable L^2-valued function, so from (15) we get that

$$||u(t)||_2 \le ||u(o)||_2 + \sqrt{2E(o)}\ t$$

Thus,

$$||\phi(t)||^2 = \int_{R^3} \{|\nabla u|^2 + m_o^2|u|^2 + |u_t|^2\}\ dx$$

is apriori bounded on finite intervals. By Theorem 2, we therefore get global existence. Thus, all the statements of Theorem 3 hold in the case $m = o$.

The case $m = o$ is often treated separately in the literature, both because $(-\Delta + m^2)^{-1}$ is unbounded in that case causing some technical difficulties and because the scattering theory is different since the rates of decay (in the sup norm) of solution of the free equation are different. Because of the above device (see Strauss [37]), the two cases can be handled on an equal footing at least as far as the existence theory is concerned.

C. Other p, n, and λ

In order to discuss the equation

$$\begin{aligned} u_{tt} - \Delta u + m^2 u &= -\lambda|u|^{p-1}u \qquad x \in R^n \\ u(x,o) &= f(x) \qquad (1) \\ u_t(x,o) &= g(x) \end{aligned}$$

for various p and n, we state the following special cases of the Sobolev estimates. The proofs generally consist of the same sort of tricks we used above in lemma 1 plus appropriate use of interpolation theorems. See, for example Friedmann [10].

Theorem 4 (Sobolev Estimates) Let m and n be positive integers, $o \leq a \leq 1$, $1 \leq p \leq \infty$, and suppose that

$$\frac{1}{p} = \frac{1}{2} - \frac{am}{n}$$

Then, for all $u \in C_0^\infty(R^n)$, there is a constant K so that

$$||u||_p \leq K||D^m u||_2^a ||u||_2^{1-a} \tag{16}$$

except when both of the following hold: $m - n/2$ is a non-negative integer and $a = 1$. In the above,

$$||D^m u||_2^2 \quad \text{denotes} \quad \sum_{j=1}^{n} || \left(\frac{\partial}{\partial x_j}\right)^m u||_2^2$$

Let us see when these estimates permit us to carry through the same analysis as in part A. Let

$$\mathcal{H}_o = D(B) \oplus L^2(R^n)$$

with A and B defined as before. Now, $J(\phi) = \langle o, -\lambda|u|^{p-1}u\rangle$ so to carry through our same methods we need an estimate of the form

$$||u||_{2p} \leq K||Bu||_2 \tag{17}$$

As indicated in part A, we can always estimate $||Du||_2$ and $||u||_2$ by $||Bu||_2$. So it follows from (16) that (17) will hold in the following cases:

$$\begin{aligned} &\underline{n=1}\ , \quad 2 \leq p \leq \infty \\ &\underline{n=2}\ , \quad 2 \leq p \leq \infty \\ &\underline{n=3}\ , \quad 2 \leq p \leq 3 \\ &\underline{n=4}\ , \quad p = 2 \end{aligned} \tag{18}$$

In all of these cases the technical details involved in proving the hypotheses of Theorem 1 are the same as in part A, and for nice initial data $\phi_0 \in D(A)$, we have a conserved energy:

$$E(o) = E(t) = \frac{1}{2}\int_{R^n} (|\nabla u|^2 + m^2|u|^2 + |u_t|^2)dx + \frac{\lambda}{p+1}\int_{R^n} |u|^{p+1}\, dx$$

Thus, if $\lambda > o$ we get an apriori bound on $||\phi(t)||$. Since the stronger hypothesis (H_1') holds we get global existence in that case. The mass zero case is handled as in part B. Thus, we have:

<u>Theorem 5</u> Let p and n satisfy one of the possibilities in (18). Then the conclusions of Theorem 3 hold locally in time if $\lambda < o$ and globally in time if $\lambda > o$ for the equation (1).

This theorem is reasonably satisfactory since we do not expect global existence to hold in the case $\lambda < o$, since both $||\phi(t)||^2$ and $\frac{\lambda}{p+1}\int |u|^p$ can become large and still conserve the energy. In fact, we will show explicitly later (part E) that global existence does not in a whole class of examples when $\lambda < o$.

Let us discuss for a moment the equation

$$u_{tt} - \Delta u + m^2 u = - \lambda u^p \tag{19}$$

where p is an integer. Notice, that if u is real-valued then (1) reduces to (19) if p is odd but not if p is even. Assuming u is real-valued, as it will be if the initial data are real, (19) has a conserved energy

$$E(t) = \int_{R^n} ((\nabla u)^2 + m^2 u^2 + u_t^2)dx + \frac{\lambda}{p+1}\int u^{p+1} dx$$

We can only insure that the term on the right will be positive if $\lambda > o$ <u>and</u> p is odd. So we only expect global existence when p is odd and in these cases it is given by Theorem 5. When p is even we can treat (19) by the same methods as in part A. For complex valued u there will not be a simple conserved energy but it doesn't matter since we don't expect global existence anyway. In fact in the case $m = o$, $n = 1$, it is shown in part E that global existence doesn't hold.

We now consider the problem in the cases not covered in (18). Since we do not have a Sobolev estimate of the form (17) we cannot even begin to try to prove local existence on our space $\mathcal{H}_o = D(B) \oplus L^2(R^n)$. However, we can prove strong local existence if we change the Hilbert space with which we work. As you will see shortly, this has unfortunate consequences when we try to prove global existence. Choose a positive integer k so that a Sobolev inequality of the form

$$||u||_\infty \leq K||B^k u||_2 \tag{20}$$

holds. By Theorem 4 and the techniques of lemmas 2 and 3 this can always be done; k will depend on n. Now we define

$$\mathcal{H}_k = \{ \langle u,v\rangle \mid u \in D(B^{2k+1}) \ , \ v \in D(B^{2k}) \}$$

with the norm

$$||\langle u,v\rangle||^2 = ||B^{2k+1}u||_2^2 + ||B^{2k}v||_2^2$$

We let A be the same operator as before but now

$$D(A) = \{\langle u,v\rangle \mid u \in D(B^{2k+1}), \ v \quad D(B^{2k})\}$$

A is self-adjoint and generates the same group W(t) discussed in the introduction. So, in order to get local existence we need just show that $J(\phi) = \langle o, -\lambda u^p\rangle$ and A satisfy the hypotheses of Theorem 1. In the following calculation we will treat B as though it acts on u^p by $B(u^p) = pu^{p-1}Bu$, which it doesn't. One need just use the technique of lemma 4 to do it right. Let $\phi = \langle u,v\rangle \in \mathcal{H}_k$. Then

$$||J(\phi)||_{\mathcal{H}_k} = |\lambda| \ ||B^{2k}(u^p)||_2$$

and the right side is less than or equal to a sum of terms of the form

$$K||(B^{k_1}u) \ ,,, \ (B^{k_p}u)||_2$$

where $2k = \sum_{i=1}^{p} k_i$ and the k_i are non-negative integers less than or equal to 2k. Let k_1 be the largest k_i. Then all the k_i for $i > 1$ are less than or equal to k. Thus, we can estimate:

$$K||(B^{k_1}u)\ldots(B^{k_p}u)||_2 \leq K||B^{k_1}u||_2 \prod_{i>1}^{p} ||B^{k_i}u||_\infty$$

$$\leq K||B^{2k+1}u||_2 \prod_{i>1}^{p} ||B^{k_i+k}u||_2$$

$$\leq K||B^{2k+1}u||_2^p$$

$$\leq K||\phi||^p$$

where we have used (20) in the second step. Thus,

$$||J(\phi)||_{\mathcal{H}_k} \leq K||\phi||^p_{\mathcal{H}_k}$$

The other estimates of Theorem 1 are proven in exactly the same way. Therefore, we can state

<u>Theorem 6</u> Let integers $n > 0$, $p \geq 2$ and any $m > 0$ and λ be given. Then there is an integer k so that if $f \in D((-\Delta + m^2)^{k+1})$ and $g \in ((-\Delta + m^2)^{k+\frac{1}{2}})$, then there is a $T > 0$ and a unique twice strongly differentiable (in t) function, $u(t,x)$, which satisfies

$$u_{tt}(x,t) - \Delta u(x,t) + m^2u(x,t) = -\lambda u(x,t)^p$$

$$u(x,0) = f(x)$$

$$u_t(x,0) = g(x)$$

for each $t \in (-T,T)$. Further, $u(t) \in D((-\Delta + m^2)^{k+1})$ $u_t(t) \in D((-\Delta + m^2)^{k+\frac{1}{2}})$ for each $t \in (-T,T)$.

The problem with this result is that we have now made it almost impossible for ourselves to prove global existence in the cases where one would expect it (for odd p). The difficulty is that we must show that the norm in $\mathcal{H}_k$

$$||\phi(t)||^2 = ||B^{2k+1}u||_2^2 + ||B^{2k}u_t||_2^2$$

does not go to infinity in finite time. From the energy inequality (for odd p) one only gets that

$$||Bu||_2^2 + ||u_t||_2^2$$

stays finite. One might try to use this and some higher order energy inequalities to prove that $||\phi(t)||_{\mathcal{H}_k}$ stays finite, but so far no one has been able to do this. It is known however that for odd p global weak (in the sense of distributions) solutions exist (see Section 5) and also that global solutions exist if the data is small enough. Thus, we have the following intriguing situation: existence of strong solutions locally, existence of weak solutions globally, but no strong global existence proof.

The use of the space $\mathcal{H}_k$, so called "escalated energy spaces", has been repeatedly emphasized by Chadam [3], [4], [5], [6]. In particular Chadam has used them to prove local existence for the coupled Maxwell-Dirac equations in three dimensions. We discuss this further in Section 6.

D. The sine-Gordon equation

We can easily apply the existence theory to the equation

$$u_{tt} - \Delta u + m^2 u = g \sin(\mathrm{Re}\{u\}) \tag{21}$$

when the number of space dimensions is $n = 1,2,3,$ or 4. If the initial data are real then the solution will be real and thus will satisfy

$$u_{tt} - \Delta u + m^2 u = g \sin(u) \tag{22}$$

which is known as the sine-Gordon equation. We treat the real solutions of (22) by studying (21) because sin(Re{z}) is bounded for all complex z while sin z grows exponentially in the imaginary directions. Using Re{u} instead of u does not affect the technical details since Re{u} and u have the same differentiability properties. One could also treat the real solutions of (22) by using a Hilbert space of real-valued functions. But, since most of our terminology is from the complex case, it is easier to treat (21). For ease of notation we will write

$$\mathrm{Re}\{u\} = \tilde{u}$$

from now on.

To begin with, we suppose $m > o$ and define $B = \sqrt{-\Delta + m^2}$, $\mathcal{H} = D(B) \oplus L^2(R^n)$, and

$$A = i\begin{pmatrix} o & I \\ -B^2 & o \end{pmatrix} \qquad D(A) = D(B^2) \oplus D(B)$$

just as in part A. But, now we have

$$J(\phi) = g\langle o, \sin\tilde{u}\rangle$$

The estimates of Theorem 1 are proven as follows:

$$\|J(\phi)\| = \|\sin\tilde{u}\|_2 \le \|u\|_2 \le \|\phi\| \tag{23}$$

$$\|AJ(\phi)\|^2 = \|B\sin\tilde{u}\|_2^2 = (\sin\tilde{u}, B^2\sin\tilde{u})_2$$

$$= \sum_{i=1}^{n}(\partial_i\sin\tilde{u}, \partial_i\sin\tilde{u})_2 + m^2(\sin\tilde{u}, \sin\tilde{u})_2$$

$$\le \|\nabla\tilde{u}\|_2^2 + m^2\|\tilde{u}\|_2^2$$

$$\le \|\nabla u\|_2^2 + m^2\|u\|_2^2$$

$$\le K\|Bu\|_2^2 \le K\|\phi\|^2$$

$$||J(\phi) - J(\psi)|| = ||\sin \tilde{u}_1 - \sin \tilde{u}_2||_2$$

$$\leq ||\tilde{u}_1 - \tilde{u}_2||_2$$

$$\leq K||\phi - \psi||$$

where $\psi = \langle u_1, v_1 \rangle$ and $\phi = \langle u_2, v_2 \rangle$

Finally,

$$||A(J(\phi) - J(\psi))|| = ||B(\sin \tilde{u}_1 - \sin \tilde{u}_2)||_2$$

$$= ||(B\tilde{u}_1)\cos \tilde{u}_1 - (B\tilde{u}_2)\cos \tilde{u}_2||_2$$

$$\leq ||(B\tilde{u}_1)(\tilde{u}_1 - \tilde{u}_2)||_2 + ||(B\tilde{u}_1 - B\tilde{u}_2)\cos \tilde{u}_2||_2$$

$$\leq ||B^2 u_1||_2 \; ||B(u_1 - u_2)||_2 + ||B(u_1 - u_2)||_2$$

$$\leq ||\phi - \psi|| \; ||A\phi|| + ||\phi - \psi||$$

In this last computation we have again for ease of exposition treated B as though it acts by differentiation. Note that the next to last step is the only place where we use the fact that the dimension is ≤ 4, because we needed the Sobolev inequality

$$||u_1 u_2||_2 \leq K||Bu_1||_2 ||Bu_2||_2$$

Since the extra hypothesis (H_1') of Theorem 2 holds, we need only show that $||\phi(t)||$ is bounded on finite intervals to show global existence. Unfortunately, (21) does not have a positive conserved energy (for all g,m). But the fact that the non-linearity is mild allows us to show apriori boundedness of $||\phi(t)||$ anyway. From the integral equation we get

$$||\phi(t)|| \leq ||e^{-iAt}\phi_0|| + ||\int_0^t e^{-iA(t-s)} J(\phi(s))ds||$$

$$\leq ||\phi_0|| + \int_0^t ||J(\phi(s))||ds$$

$$\leq ||\phi_0|| + \int_0^t ||\phi(s)||ds$$

so by iteration

$$||\phi(t)|| \leq ||\phi_0||e^t$$

Thus $||\phi(t)||$ is bounded on bounded intervals so by Theorem 2 the solution exists globally. The mass zero case is handled just as in part B. We summarize:

<u>Theorem 7</u> Let $n = 1,2,3,$ or 4 and let $m \in [0,\infty)$ and $g \in (-\infty,\infty)$ be given. Then for each $f \in D(-\Delta + m^2)$, $g \in D((-\Delta + m^2)^{1/2})$ the initial value problem

$$u_{tt} - \Delta u + m^2 u = g \sin u$$

$$u(x,o) = f(x)$$

$$u_t(x,o) = g(x)$$

has a unique global solution u such that u is twice continuously differentiable as an $L^2(R^n)$-valued function of t, $u \in D(-\Delta + m^2)$ and $u_t \in D((-\Delta + m^2)^{\frac{1}{2}})$ for each t.

Notice that if we just want to solve the integral equation

$$\phi(t) = e^{-iAt} + \int_0^t e^{-iA(t-s)} J(\phi(s))ds$$

instead of

$$\phi'(t) = - iA\phi(t) + J(\phi(t))$$

then, by Corollary 1 of Theorem 1, we only need the hypothesis

$$(H_o^L) \qquad ||J(\phi) - J(\psi)|| \leq C(||\phi||,||\psi||)\ ||\phi - \psi||$$

Since this hypothesis holds for the sine - Gordon equation in all dimensions (we only used the Sobolev inequality in our estimate for $||A(J(\phi) - J(\psi))||$) we conclude that (21) has weak global solutions in all dimensions. By weak, we mean here that the corresponding integral equation has a continuous $\mathcal{H}$-valued global solution.

E. <u>An example when global existence fails</u>

Let us consider the equation

$$(24) \qquad u_{tt} - u_{xx} = u^p \qquad x \in R$$

$$u(x,o) = u_o(x)$$

$$u_t(x,o) = v_o(x)$$

where p is an integer ≥ 2. We know that local existence holds and, anticipating Section 3, we can guarantee that if the solution u starts out in C_o^∞ (i.e. $u_o \in C_o^\infty(R)$, $v_o \in C_o^\infty(R)$) then on some finite t interval the solution will be twice continuously differentiable. Further, using another result of section 4, it will have compact support. These regularity statements don't affect the ideas below, they just allow us to integrate by parts with impunity. If u_o and v_o are real-valued then u will be real-valued and

$$E(t) = \frac{1}{2}\int_R \{(\nabla u)^2 + m^2u^2 + u_t^2 - \frac{2}{p+1}u^{p+1}\}\,dx$$

is the conserved energy. Thus we do not expect global existence for any $p \geq 2$. We will show that if u_o and v_o are chosen correctly, then

$$F(t) = \int_R u(x,t)^2 dx$$

goes to infinity in finite time. Suppose that we can find an $\alpha > 0$ and initial data u_0 and v_0 so that

(A) $(F(t)^{-\alpha})'' \leq 0$ for all $t \geq 0$

(B) $(F(t)^{-\alpha})' < 0$ at $t = 0$

Then $F(t)^{-\alpha}$ will go to zero in finite time; see the Figure. Condition (B) is automatically satisfied by choosing u_0 and v_0 to have the same sign on $(-\infty,\infty)$ since

$$(F(0)^{-\alpha})' = -\alpha F(0)^{-1-\alpha}F'(0) = -2\alpha F(0)^{-1-\alpha}\int u_0 v_0 dx$$

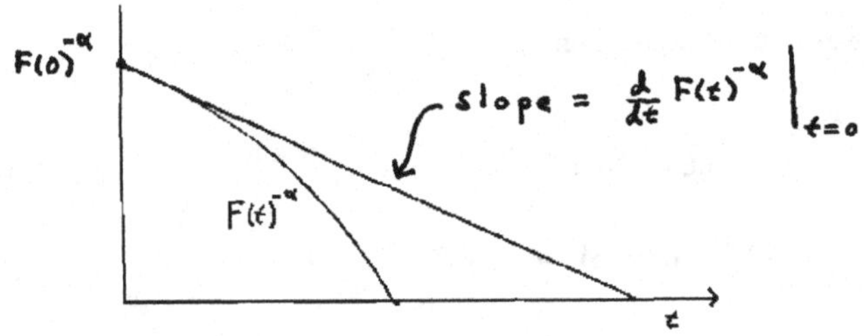

It remains to arrange for (A) to hold. Since $F(t) \geq 0$ this is the same showing that $Q(t) \geq 0$ where

$$Q(t) \equiv (-\alpha)^{-1}F^{\alpha+2}(F^{-\alpha})'' = F''F - (\alpha+1)(F')^2$$

But,

$$F'(t) = 2\int uu_t dx$$

$$F''(t) = 2\int (uu_{tt} + u_t^2)dx$$

$$= 4(\alpha + 1)\int u_t^2 dx + 2\int (uu_{tt} - (2\alpha + 1)u_t^2)dx$$

so,

$$Q(t) = 4(\alpha + 1)\left\{\left(\int u^2 dx\right)\left(\int u_t^2 dx\right) - \left(\int uu_t dx\right)^2\right\}$$

$$+ 2F(t)\left\{\int uu_{tt}dx - \int(2\alpha + 1)u_t^2dx\right\}$$

The first term on the right is positive by the Schwarz inequality, so we need only arrange that $H(t) \geq 0$ where

$$H(t) \equiv \int uu_{tt}dx - (2\alpha + 1)\int u_t^2dx$$

$$= \int u^{n+1}dx + \int uu_{tt}dx - (2\alpha + 1)\int u_t^2dx$$

$$= \int u^{n+1}dx - \int u_x^2dx - (2\alpha + 1)\int u_t^2dx$$

The conserved energy for (24) is

$$E(t) = \frac{1}{2}\int(u_x^2 + u_t^2)dx - \frac{1}{p+1}\int u^{p+1}dx$$

That is, $E(t)$ is independent of t. Thus, if we choose α so that $2(2\alpha + 1) = p + 1$, we have

$$H(t) = -(p+1)E(t) + 2\alpha\int u_x^2dx \tag{24}$$

$$= -(p+1)E(0) + 2\alpha\int u_x^2dx$$

Thus, if $E(0) < 0$, then H is always strictly positive since $\alpha = \frac{1}{4}(n-1) \geq 0$. Now, choosing $u_0 \geq 0$, $v_0 \geq 0$ so that (B) is satisfied we scale u_0 by multiplying by a positive constant until $E(0) < 0$ (this will eventually happen since $p+1 > 2$). For any such initial data, $F(t)$ goes to infinity in finite time.

If we consider instead the differential equation

$$u_{tt} - u_{xx} = -u^p$$

then $H(t)$ again satisfies (24), but now the conserved energy is

$$E(t) = \frac{1}{2}\int (u_x^2 + u_t^2)dx + \frac{1}{p+1}\int u^{p+1}dx$$

If p is even then by choosing $u_o(x) \leq 0$, $v_o(x) \leq 0$ (thus satisfying (B)) with u_o sufficiently large we can obtain $E(o) < 0$ and thus the solution blows up in finite time. If, on the other hand, p is odd, then $E(t)$ is always greater than or equal to zero so the above argument does not work. This is not surprising since we have proven global existence in this case in parts A, B, and C. Notice that we always had to choose the initial data large in order to get the solution to blow up. Later we will see that if the initial data are small enough then global solutions exist independent of whether p is even or odd or the sign of λ.

The author learned this simple example from H.Levine [18]. The details are taken from Reed-Simon [27]. Examples of non-existence of global solutions have been known for a long time. See for example Keller [16] or Glassey [11].

F. The Coupled Dirac and Klein-Gordon Equations

In part D we saw that if the non-linearity is "mild" in an appropriate sense, then we can get global existence even though the conserved energy is not positive. A much more striking example of global existence is Chadam's proof [5] of global existence for the Yukawa coupled Dirac and Klein-Gordon equations in one-dimension. Let α and β be two by two Hermetian matrices which satisfy

$$\alpha^2 = I = \beta^2 , \quad \beta\alpha + \alpha\beta = 0$$

Then we can write the free Dirac equation (for one space dimension) as

$$\text{(F 1)} \qquad \frac{d}{dt}\,\psi(t) = -i\left(i\alpha\frac{\partial}{\partial x} - m_e\beta\right)\psi(t)$$

where $\psi(t) = \langle\psi_1(x,t), \psi_2(x,t)\rangle$ is a C^2-valued function on R^2. We would like to solve the coupled system

$$\text{(F 2)} \qquad \frac{d}{dt}\psi(t) = -i\left(i\alpha\frac{\partial}{\partial x} - m_e\beta\right)\psi(t) - ig\beta u(t)\psi(t)$$

$$\text{(F 3)} \qquad u_{tt} - u_{xx} + m_o^2 u = \overline{\psi}(t)\cdot\beta\psi(t)$$

where g is a real coupling constant, $u(t) = u(x,t)$ is an R-valued function on R^2, and $\overline{\psi}(t)\cdot\beta\psi(t)$ denotes the dot product in C^2 of $\psi(t)$ with $\beta\psi(t)$. These equations have a conserved energy but it is not bounded below and is therefore not much use. So we don't have anything to lose by letting $u(t)$ be complex-valued, defining

$$\tilde{u}(t) = \mathrm{Re}(u(t))$$

and rewriting (F 2) with $\tilde{u}(t)$ on the right side instead of $u(t)$. Since β is Hermetian $u(t)$ will remain real if its initial data are real. As usual we rewrite (F3) as a first order system for $\phi(t) = \langle u(t), v(t)\rangle$

$$\text{(F 4)} \qquad \frac{d}{dt}\,\phi(t) = -iA_o\phi(t) + J_o(\psi(t))$$

where $B_o = \sqrt{-\Delta+m_o^2}$, $J_o(\psi(t)) = \langle o, \overline{\psi}(t) \cdot \beta\psi(t)\rangle$ and,

$$A_o = \begin{pmatrix} o & I \\ -B_o^2 & o \end{pmatrix}$$

Letting $D_e = i\alpha\frac{\partial}{\partial x} - m_e\beta$ and combining (F2) and (F4) we have the coupled first order system:

$$\psi'(t) = -iD_e\psi(t) + J_e(\phi(t))$$

(F 5)

$$\phi'(t) = -iA_o\phi(t) + J_o(\psi(t))$$

where $$J_e(\phi(t)) = -ig\beta\tilde{u}(t)\psi(t)$$

Finally, if we set

$$\Xi(t) = \langle\psi(t), \phi(t)\rangle$$

$$J(\Xi(t)) = \langle J_e(\phi(t)), J_o(\psi(t))\rangle$$

then we can write (F5) as

(F 6) $$\Xi'(t) = -iA\Xi(t) + J(\Xi(t))$$

where A is the four by four operator matrix

$$A = \begin{pmatrix} D_e & o \\ o & A_o \end{pmatrix}$$

Thus we have rewritten our coupled equations in a form so that we can apply the existence theory of Section 1. Let $B_e = \sqrt{-\Delta+ m_e^2}$. Then we take as our Hilbert space

$$\mathcal{H} = D(B_e) \oplus D(B_e) \oplus D(B_o) \oplus L^2(R)$$

where the norm of $\Xi = \langle\psi_1,\psi_2, u,v\rangle$ is

$$||\Xi||^2 = ||B_e\psi_1||_2^2 + ||B_e\psi_2||_2^2 + ||B_o u||_2^2 + ||v||_2^2$$

A is self-adjoint on

$$D(A) = D(B_e^2) \oplus D(B_e^2) \oplus D(B_o^2) \oplus D(B_o)$$

To see that J is a well-defined mapping on $\mathcal{H}$ we use the Sobolev inequality

$$\text{(F 7)} \qquad ||f(x)||_\infty \leq c||f'(x)||_2^{1/2} \; ||f(x)||_2^{1/2}$$

valid in dimension $n = 1$ (see part C). This implies

$$||f||_\infty \leq c||B_e f||_2^{1/2} \; ||f||_2^{1/2}$$

and

$$||f||_\infty \leq c||B_o f||_2^{1/2} \; ||f||_2^{1/2}$$

Now $||J(\Xi(t))||^2$ has terms of the form $||B_e(\tilde{u}\psi_i)||_2^2$ and of the form $||\overline{\Psi}\cdot\beta\Psi||_2^2$ and we can estimate (treating the B's like differentiation as usual)

$$||B_e(\tilde{u}\psi_i)||_2 \leq ||\psi_i||_\infty \; ||B_e\tilde{u}||_2 + ||B_e\psi_i||_2||\tilde{u}||_\infty$$

$$\leq c||B_e\psi_i||_2 \; ||B_o\tilde{u}||_2 + ||B_e\psi_e||_2||B_o\tilde{u}||_2$$

$$\leq c||\Xi||^2$$

and similarly for the other term. We have used the fact that $B_e^{-1}B_o$ and $B_o^{-1}B_e$ are bounded (by the functional calculus). In fact, using this same idea one can easily check that J satisfies hypotheses (H_m') and (H_m^L) for $m = o,1$. Thus we get a local solution of (F6) by Theorem 1.

Our problem is to get a global solution. Since (H_1') is satisfied we need just show that $||\Xi(t)||$ is apriori bounded. But we cannot use the conserved energy to do this since it is not bounded below so our only hope is to use the integral equations directly (as we did for the sine-Gordon equation). The two integral equations are

$$\psi(t) = e^{-iD_e t}\psi(o) + \int_o^t e^{-iD_e(t-s)} J_e(\phi(s))ds$$

$$\phi(t) = e^{-iA_o t}\phi(o) + \int_o^t e^{-iA_o(t-s)} J_o(\psi(s))ds$$

so

$$||\psi(t)||_e \leq c_e + \int_o^t ||J_e\phi(s)||_e ds$$

$$||\phi(t)||_o \leq c_o + \int_o^t ||J_o(\psi(s))||_o ds \tag{F 8}$$

where the $||\cdot||_e$ and $||\cdot||_o$ norms are defined by

$$||\Xi||^2 = ||\psi||_e^2 + ||\phi||_o^2$$

$$||\psi||_e^2 = ||B_e\psi_1||_2^2 + ||B_e\psi_2||_2^2$$

$$||\phi||_o^2 = ||B_o u||_2^2 + ||v||_2^2$$

Using (F7), we have

$$||J_e(\phi(s))||_e \leq c\sum_{i=1}^{2} ||B_e(\tilde{u}\psi_i)||_2$$

$$\leq c\sum_{i=1}^{2}\{||B_o\tilde{u}||_2||B_e\psi_i||_2^{1/2}||\psi_i||_2^{1/2} + ||B_o\tilde{u}||_2^{1/2}||\tilde{u}||_2^{1/2}||B_e\psi_i||_2\} \tag{F 9}$$

and

$$||J_o(\psi(s))||_o = ||\overline{\psi}\cdot\beta\psi||_2$$

$$\leq c\{||\psi_1^2||_2 + ||\psi_2^2||_2\}$$

$$\leq c\{||B_e\psi_1||_2^{1/2}||\psi_1||_2^{3/2} + ||B_e\psi_2||_2^{1/2}||\psi_2||_2^{3/2}\} \tag{F 10}$$

So far, the situation looks pretty hopeless since the right side of (F9) and (F10) are quadratic expressions in the norms $||\cdot||_e$ and $||\cdot||_o$ so we can't expect to get anywhere with (F8) by interation. In fact, we can't get anywhere unless we use more information about the coupled system (F2), (F3). So far, the only property of the matrices α and β which we have used is that they are Hermetian (so that D_e is Hermetian). Using the other properties of α and β one can easily show (see for example Bjorken and Drell [1] or Chadam [5]) that the total electric charge

$$Q(t) = \int_{-\infty}^{\infty} |\psi_1(x,t)|^2 dx + \int_{-\infty}^{\infty} |\psi_2(x,t)|^2 dx$$

is a conserved quantity; $Q(t) = Q$. This is an immediate help for (F10) can now be written

$$\text{(F 11)} \qquad ||J_o(\psi(s))||_o \leq c\,\{||B_e\psi_1||_2^{1/2} + ||B_e\psi_2||_2^{1/2}\}$$

$$\leq c||\psi(s)||_e^{1/2}$$

However, conservation of charge as it stands is only a small help for (F9) which can now be written

$$\text{(F 12)} \quad ||J_e(\phi(s))||_e \leq c \sum_{i=1}^{2} \{||B_o\tilde{u}||_2 ||B_e\psi_i||_2^{1/2} + ||\tilde{u}(t)||_\infty ||B_e\psi_i||\}$$

Our only hope to make this better is to get an estimate on $||\tilde{u}(t)||_\infty$. Fortunately, we can do this by using special properties of the propagator and conservation of charge(again!). $u(t)$ satisfies the equation

$$u(t) = u_o(t) + \int_0^t [B_o^{-1} \sin B_o t]\, J_o(\psi(s))ds$$

where $u_o(t)$ is the first component of $e^{-iA_o t}\phi(o)$. Since the free propagation preserves the norm $||B_o u||_2^2 + ||u_t||_2^2$ we have

$$||u_o(t)||_\infty \leq c||B_o u_o(t)||_2^{1/2} ||u_o(t)||_2^{1/2} \leq c_1$$

for all t. Thus,

(F 13) $||u(t)||_\infty \leq c_1 + \int_0^t ||[B_0^{-1} \sin B_0(t-s)] \, J_0(\psi(s))||_\infty ds$

Now, $B_0^{-1}\sin B_0 t$ acts by convolution by a function $H(x,t-s)$ which is the inverse Fourier transform of $(k^2+m_0^2)^{-1/2} \sin(\sqrt{k^2+m_0^2}(t-s))$ Since $J_0(\psi(s))$ is quadratic in the $\psi_i(s)$ and charge is conserved, the L^1 norm of $J_0(\psi(s))$ is uniformly bounded independently of s. Thus, if $||H(x,t)||_\infty \leq c$ we can conclude that

(F 14) $$||(B_0^{-1} \sin B_0(t-s))J_0(\psi(s))||_\infty \leq c$$

for all s and t. One can compute the inverse Fourier transform and use properties of Bessel functions to conclude that $||H(x,t)||_\infty \leq c$ but the easiest thing to do is to notice that the inverse Fourier transform of $\frac{\sin kt}{k}$ is uniformly bounded for all t since it is just a fixed constant times the characteristic function of an interval. On the other hand

$$\frac{\sin t\sqrt{k^2+m^2}}{\sqrt{k^2+m^2}} - \frac{\sin t k}{k}$$

is an L^1 function whose L^1 norm is uniformly bounded on any finite t interval. Thus

$$||H(x,t-s)||_\infty \leq c$$

for s and t in an finite interval. Choose such an interval $[-T,T]$. Then for $t \in [-T,T]$, $||\tilde{u}(t)||_\infty \leq c$, so (F12) becomes

(F 15) $||J_e(\phi(s))||_e \leq c\{||\phi(s)||_0 ||\psi(s)||_e^{1/2} + ||\psi(s)||_e\}$

Therefore, if we set

$$f(t) = ||\phi(t)||_0$$

$$g(t) = ||\psi(s)||_e^{1/2}$$

then (F8) becomes the following set of coupled inequalities

(F 16a) $$g(t)^2 \le c_e + a \int_0^t (f(s)g(s) + g(s)^2)ds$$

(F 16b) $$f(t) \le c_0 + b \int_0^t g(s)ds$$

where a can depend on T but not on t. Substituting the second inequality into the first, we have

$$g(t)^2 \le c_e + ac_0 \int_0^t g(s)ds + a \int_0^t g(s)^2 ds + ab \int_0^t g(s) \int_0^s g(r)dr$$

Using the Schwartz inequality on the second and fourth terms, we have

$$g(t)^2 \le c_e + d \int_0^t g(s)^2 ds$$

where d depends on T but not on t. Therefore, by iteration $g(t)^2$ is bounded on $[-T,T]$. Substituting back into (F16b) we conclude that f(t) is also bounded on $[-T,T]$. Since T was arbitrary, we have proven that the solution $\Xi(t)$ of (F6) is apriori bounded on any finite interval and therefore exists globally by Theorem 2.

This proof of global existence in one dimension is due to Chadam [5] where the method is applied to the coupled Maxwell-Dirac equations. Using the escalated energy spaces introduced in part c, one can easily show the existence of <u>local</u> strong solutions in three dimensions (Chadam[4]). However, the question of global existence in three dimensions is open. For more discussion, see Section 6.

3. Smoothness of Solutions

As we have remarked before, the existence theorems in Section 1 are not completely satisfactory from a classical point of view. For example we would like to know that if we choose the data at time zero to be smooth (say C^∞), then the solution of (1) will stay smooth and satisfy the equation in the classical sense. Essentially two kinds of further hypotheses are needed. In our examples the powers of A, where

$$A = i\begin{pmatrix} 0 & I \\ \Delta - m^2 & 0 \end{pmatrix}$$

act like powers of the Laplacian. So, if the solution $\phi(t)$ remains in the domain of high enough powers of A then we should get smoothness in the x variables. To achieve this we require only higher order estimates of the same kind as (H_1), (H_1^L) in Theorem 1. Secondly, looking at our equation

$$\phi'(t) = -iA\phi(t) + J(\phi(t)) \tag{6}$$

we can see that to get higher differentiability of ϕ in t, we will have to assume some "differentiability" in J. Further, the derivatives of $\phi(t)$ must remain in the domain of appropriate powers of A. These factors make our conditions on J awkward to formulate; but as you will see, it works out quite easily in applications. As in Theorem 1, the hypotheses (H_i) in part (a) below follow from (H_i^L). We state them thus for easy comparison with Theorem 9. We will refer to the hypothesis in part b below as "condition J_m".

<u>Theorem 8</u> (local smoothness)
(a) Let A be a self-adjoint operator on a Hilbert space $\mathcal{H}$ and J a mapping which takes $D(A^j)$ into $D(A^j)$ for all $1 \le j \le m$ and which satisfies (for $j = 0,1,\ldots,m$)

(H_j) $\quad ||A^j J(\phi)|| \le C(||\phi||,\ldots,||A^j\phi||)\,||A^j\phi||$

(H_j^L) $\quad ||A^j(J(\phi) - J(\psi))||$

$$\le C(||\phi||,||\psi||,\ldots,||A^j\phi||,||A^j\psi||)\,||A^j\phi - A^j\psi||$$

for all $\phi,\psi \in D(A^j)$ where each constant C is a (everywhere finite) function of all its variables. Then for each $\phi_o \in D(A^m)$, $m \geq 1$, there is a T_m so that (6) has a unique solution $\phi(t)$ for $t \in [0,T_m)$ with $\phi(t) \in D(A^m)$ for all $t \in [0,T_m)$. For each set of the form $\{\phi \mid ||A^j\phi|| \leq a_j,\ j = 0,\ldots,m\}$, T can be chosen uniformly for ϕ_o in the set.

(b) In addition to the hypotheses in (a) assume that for each $j \leq m$, J has the following property: If a solution ϕ is j times strongly continuously differentiable with $\phi^{(k)}(t) \in D(A^{m-k})$ and $A^{m-k}\phi^{(k)}(t)$ is continuous for all $k \leq j$, then $J(\phi(t))$ is j times differentiable, $d^jJ(\phi(t))/dt^j \in D(A^{m-j-1})$, and $A^{m-j-1}d^jJ(\phi(t))/dt^j$ is continuous. Then the solution given in part (a) is m times strongly differentiable in t and $d^j\phi(t)/dt^j \in D(A^{m-j})$.

Proof The proof of part (a) is essentially the same as the proof of Theorem 1 except that we take $X(T_m,\alpha,\phi_o)$ to be the set of functions $\phi(\cdot)$ on $[0,T_m)$ so that $\phi(t),\ldots,A^m\phi(t)$ are strongly continuous and

$$\sum_{j=0}^{m} \sup_{t\in[0,T)} ||A^j\phi(t) - e^{-iAt}A^j\phi_o|| \leq \alpha$$

Then one proves that S is a contraction as before.

Part (b) is proven by induction. We know from part (a) that $\phi(t)$ is strongly continuously differentiable and $\phi'(t) = -iA\phi(t) + J(\phi(t))$. By the same arguments as in Theorem 1,

$$A\phi(t) = Ae^{-iAt}\phi_o + A\int_0^t e^{-iA(t-s)}J(\phi(s))ds$$

$$= e^{-iAt}A\phi_o + \int_0^t e^{-iA(t-s)}AJ(\phi(s))ds$$

and from this it follows (using another argument in Theorem 1: see (10)) that $A\phi(t)$ is strongly continuously differentiable. Therefore by the hypotheses on J, $J(\phi(t))$ is strongly continuously differentiable, $dJ(\phi(t))/dt \in D(A^{m-2})$, and $A^{m-2}dJ(\phi(t))/dt$ is continuous. Thus, $\phi'(t)$ is strongly differentiable,

$$\phi''(t) = -A\phi'(t) + \frac{d}{dt}J(\phi(t))$$

$$= (-iA)^2\phi(t) - iAJ(\phi(t)) + \frac{d}{dt} J(\phi(t))$$

$\phi''(t) \in D(A^{m-2})$, and $A^{m-2}\phi''(t)$ is continuous. We now repeat the argument again $(dJ(\phi(t))/dt$ is differentiable by hypothesis since we now know that $\phi(t)$ is twice continuously differentiable) to conclude that $\phi(t)$ is three times strongly differentiable and so forth. ∎

Notice that the interval on which the solution exists depends on m. In particular, T_m may go to zero as $m \longrightarrow \infty$. As in Section 1 if we have slightly stronger estimates and apriori boundedness of $||\phi(t)||$, then we get global existence and smoothness.

<u>Theorem 9</u> (global existence and smoothness) Let A, J, and $\mathcal{H}$ satisfy the hypotheses of part (a) of Theorem 8 except that for $j = 1,2,\dots,m$ (H_j) is replaced by

(H_j')
$$||A^jJ(\phi)|| \leq C(||\phi||,\dots,||A^{j-1}\phi||)||A^j\phi||$$

Let $\phi_o \in D(A^m)$ and suppose that on any finite interval of existence of the solution $\phi(t)$, $||\phi(t)||$ is bounded. Then the solution $\phi(t)$ of (6) is global in t. Further, if J satisfies condition J_m then $\phi(t)$ is m times strongly differentiable for all t and satisfies

$$\frac{d^j}{dt^j}\phi(t) \in D(A^{m-j})$$

<u>Proof</u> The idea is the same as in Theorem 2. On any finite interval of existence $||\phi(t)||$ is apriori bounded. On this same interval we showed in Theorem 2 that $||A\phi(t)|| \leq ||A\phi_o||\, e^{K_1 t}$. Now for $A^2\phi(t)$ we have the estimate

$$||A^2\phi(t)|| \leq ||A^2\phi_o|| + \int_o^t ||A^2J(\phi(s))||ds$$

$$\leq ||A^2\phi_o|| + \int_o^t C(||\phi(s)||,||A\phi(s)||)||A^2\phi(s)||ds$$

by (H_2'). Since we already have $||\phi(s)||$ and $||A\phi(s)||$ apriori bounded, $C(||\phi(s)||,||A\phi(s)||)$ is less than some constant K_2 on the interval in question. Thus,

$$||A^2\phi(t)|| \leq ||A^2\phi_o|| + K_2\int_o^t ||A^2\phi(s)||ds$$

so,

$$||A^2\phi(t)|| \leq ||A^2\phi_o||e^{K_2 t}$$

Now we have $||A^2\phi(t)||$ bounded so we can continue in the same fashion to conclude that $||A^j\phi(t)||$ is apriori bounded on any finite interval of existence for $j = 1,\ldots,m$.

The length of the interval T_m in Theorem 6 on which the solution exists depends only on the constants $C(||\phi_o|| + \alpha,\ldots,||A^m\phi_o|| + \alpha)$. Therefore, as in Theorem 2 we can extend the solution past the end of any finite interval. So, the solution is global in t. The other statements of the theorem follow from Theorem 6. ∎

Corollary Let A, J, and $\mathcal{H}$ satisfy the hypotheses of Theorem 9 for each $m = 1,2,\ldots$ And suppose that J satisfies condition J_m for each m. Then for each $\phi_o \in \bigcap_{j=1}^{\infty} D(A^j)$, the equation (6) has a unique global solution so that $\phi(t)$ is infinitely often strongly differentiable and each derivative is in $\bigcap_{j=1}^{\infty} D(A^j)$.

Example 1 First we will apply these ideas to

$$\begin{aligned} u_{tt} - \Delta u + m^2 &= - u^3 \\ u(x,o) &= f(x) \\ u_t(x,o) &= g(x) \end{aligned} \tag{25}$$

We use exactly the same set up as in part A of Section 2. We need just verify the higher order estimates and the hypotheses on J in part b of Theorem 6. The higher order estimates are easy and use exactly the same techniques (i.e. the same Sobolev estimate) as in part A. For example:

$$||A^m J(\phi)|| = ||B^m u^3||_2$$

The term on the right is less than or equal to a sum of terms of the form $||(B^{m_1}u)(B^{m_2}u)(B^{m_3}u)||_2$ with $m_1 + m_2 + m_3 = m$. Let m_1 be the largest of the m_i. Then,

$$||(B^{m_1}u)(B^{m_2}u)(B^{m_3}u)||_2 \leq C||B^{m_1+1}u||_2||B^{m_2+1}u||_2||B^{m_3+1}u||_2$$

$$\leq C_1||B^{m+1}u||_2||B^{m}u||_2||B^{m}u||_2$$

$$\leq C_1||A^{m-1}\phi||^2||A^{m}\phi||$$

so the hypotheses (H'_m) are satisfied for all m. (Note that we are again treating B as though it were just differentiation - see Section 2, part D). The proof of (H^L_m) is similar.

Checking condition J_m is also easy. To see what is involved let us take m = 2. Suppose that $\phi(t) = \langle u(t), u_t(t)\rangle$ is the solution of part a of Theorem 6 in our case and that $\phi(t) \in D(A^2), \phi'(t) \in D(A)$ and $A\phi'(t)$ is continuous in t. We must show that $J(\phi(t))$ is strongly differentiable. Since $\phi(t) \in D(A^2)$, $u(t) \in D(B^3)$ and $u_t(t) \in D(B^2)$. The difference quotient for J,

$$\frac{1}{h}(J(\phi(t+h)) - J(\phi(t))) = \langle o, -\frac{u(t+h)^3 - u(t)^3}{h}\rangle$$

can be written as the sum of three terms one of which is

$$\langle o, u(t)^2\left(\frac{u(t+h) - u(t)}{h}\right)\rangle$$

Thus, by the same Sobolev inequality we have the estimate:

$$||u(t)^2\left\{\left(\frac{u(t+h) - u(t)}{h}\right) - u_t(t)\right\}||_2$$

$$\leq C||Bu(t)||_2^2\ ||B\left[\frac{u(t+h) - u(t)}{h} - u'(t)\right]||_2$$

Since $\phi(t)$ is strongly differentiable, the right hand side converges to zero. The same argument works for the other terms, so we conclude that $J(\phi(t))$ is strongly differentiable,

$$\frac{d}{dt} J(\phi(t)) = \langle o, -3u(t)^2 u_t(t)\rangle$$

so $\frac{d}{dt} J(\phi(t)) \in D(A^{2-1-1}) = \mathcal{H}$ and is continuous. All other cases are essentially identical.

Therefore, the corollary to Theorem 7 is applicable. Let us suppose that

$$f(x) \in C_o^\infty(R^3) \quad , \quad g(x) \in C_o^\infty(R^3)$$

Then $\phi_o = \langle f,g\rangle \in \bigcap_{j=1}^{\infty} D(A^j)$ so the solution $\phi(t) = \langle u(x,t), u_t(x,t)\rangle$ has the property that $u(t,x)$ is an infinitely often strongly differentiable $L^2(R^3)$-valued function of t and each derivative is in $\bigcap_{n=1}^{\infty} D(B^n)$ for each t. The L^2-derivatives of $u(t,x)$ in the time direction are just the weak time derivatives (i.e. derivatives in the sense of distributions) of $u(t,x)$. Further if $D_t^m u(t,x)$ is one of these derivatives then since $D_t^m u(t,x) \in \bigcap_{j=1}^{\infty} D(B^j)$ we have $\Delta^p D_t^m u(t,x) \in L^2(R^4)$ locally for all p and m. Therefore, by Sobolev's lemma (see [27], p. 52), $u(t,x)$ is a C^∞ function of x and t.

The proofs in the case $m = o$ and for other p satisfying (18) are the same. Thus, we have:

<u>Theorem 10</u> Let $m \geq o$, $\lambda > o$ and suppose n and p are in one of the cases (18) (in §2, part C) where p is an odd integer. Suppose f, g, $\in C_o^\infty(R^n)$. Then there is a unique C^∞ function $u(t,x)$ on R^4 which satisfies

$$u_{tt} - \Delta u + m^2 u = -\lambda u^p$$

$$u(x,o) = f(x)$$

$$u_t(x,o) = g(x)$$

We remark that various bounds on the growth (in time) of the L^2-norms of $u(t,x)$ and its derivatives follow from our estimate and the energy inequality.

Example 2 For the cases of high p in dimension $n \geq 3$ and for p even or the case where the non linear term is $-\lambda u^p$ with $\lambda < o$ we cannot use Theorem 9 since there is no apriori estimate on the norm $||\phi(t)||$ as discussed in Section 2 part C. (Notice that for high p when $n \geq 3$, we mean the norm in an escalated energy space). Nevertheless the hypotheses of Theorem 6 are satisfied for each m as can be easily checked by making similar calculations to those in Example 1 and in part C of Section 2. Thus we can apply Theorem 6 and Sobolev's lemma as above to conclude that if the initial data is smooth enough, then the solution continues to have the same degree of smoothness in the interval where it exists and satisfies the appropriate differential equation in a classical sense. One cannot get that C^∞ data go into C^∞ data in these cases because the higher m, the shorter the interval of existence $(-T_m, T_m)$ may become.

Example 3 (sine-Gordon equation)

For the sine-Gordon equation the situation is a little different in that for high m the term.

$$||A^m J(\phi)|| = ||B^m \sin \tilde{u}||_2$$

will have more and more products of $(B^{m_i}u)$ multiplied together. This poses no difficulty in one or two diminsions since in those cases $||u||_p \leq C||Bu||_2$ for all $p < \infty$. But in three dimensions this only holds for $p \leq 6$ and in four dimensions for $p \leq 4$. However we can also use the estimate

$$||B^{n_o}u||_\infty \leq C||B^{n_o+3}u||_2 \tag{26}$$

which is valid for $n \leq 4$ (see the Sobolev inequalities in Section 2, part C). In the following we will use just (26) and

$$||B^{n_o}u||_4 \leq C||B^{n_o+1}u||_2 \tag{27}$$

Thus the proof we give simultaneously handles the cases $n = 1,2,3,4$. We have already proven the estimates of Theorem 7 in Section 2, part D for the cases $m = o,1$. For a general m,

$$||A^m(J(\phi) - J(\psi))|| = ||B^m(\sin \tilde{u}_1 - \sin \tilde{u}_2)||_2 \tag{28}$$

We will show that the term on the right side is less than or equal to

$$C(||B^m\tilde{u}_1||_2, ||B^m\tilde{u}_2||_2)||B^{m+1}(\tilde{u}_1 - \tilde{u}_2)||_2$$
$$\leq C(||A^{m-1}\phi||, ||A^{m-1}\psi||)\ ||A^m(\phi - \psi)|| \tag{29}$$

This will prove that the hypotheses (H_m^L) and (H_m') (taking $\psi = o$) hold for all m. Let $\phi = \langle u_1, v_1 \rangle$, $\psi = \langle u_2, v_2 \rangle$. First we write out (28) treating B like differentiation. Adding and subtracting the same term a number of times and using the estimates:

$$|\sin \tilde{u}_1 - \sin \tilde{u}_2| \leq |\tilde{u}_1 - \tilde{u}_2|$$

$$|\cos \tilde{u}_1 - \cos \tilde{u}_2| \leq |\tilde{u}_1 - \tilde{u}_2|$$

we find that the right hand side of (28) is less than or equal to a finite sum of terms of the form

$$||(B^{m_1}\tilde{w})\ldots(B^{m_p}\tilde{w})\ (\tilde{w})^\alpha||_2 \tag{30}$$

where $\sum_{i=1}^{p} m_i = m$, α is a zero or one, and where w stands for either $\tilde{u}_1, \tilde{u}_2$, or $\tilde{u}_1 - \tilde{u}_2$. We may always assume $m_1 = \max_j m_j$. Now, for $m \geq 5$, we must $m - m_j \geq 3$ for all m_j, $j \neq 1$. Thus, we can just use (26) to conclude

$$||(B^{m_1}\tilde{w})\ldots(B^{m_p}\tilde{w})(\tilde{w})^\alpha||_2 \leq ||B^{m_1}\tilde{w}||_2\ ||\tilde{w}||_\infty^\alpha\ ||\prod_{j=2}^{p}||B^{m_j}\tilde{w}||_\infty$$

$$\leq C\ ||B^{m+1}w||_2\ ||B^m w||_2 \prod_{j=2}^{p}||B^m w||_2$$

in case $m \geq 5$.

For $m = 2,3,4$ we just check each of the possibilities using (26), (27), or both. For example, when $m = 3$, there are three kinds of terms:

$$m_1 = 3 \qquad ||(B^3w)w||_2 \leq ||B^4w||_2||Bw||_2 \qquad \text{by (21)}$$

$$m_1 = 2 \qquad ||(B^2w)(Bw)w||_2 \leq ||(B^2w)(Bw)||_2||w||_\infty$$

$$\leq ||(B^2w)(Bw)||_2||B^3w||_2 \qquad \text{by (26)}$$

$$\leq ||B^3w||_2||B^2w||_2||B^3w||_2 \qquad \text{by (27)}$$

$$m_1 = 1 \qquad ||(Bw)(Bw)(Bw)w||_2 \leq ||(Bw)^2||_2||Bw||_\infty||w||_\infty$$

$$\leq K||(Bw)^2||_2||B^4w||_2||B^3w||_2$$

$$\leq K||B^2w||_2^2||B^4w||_2||B^3w||_2$$

The cases $m = 2$ and $m = 4$ are checked similarly. Notice that on the right hand side of all of these estimates there is a most one power of a term with B^{n+1}. For this reason the stronger estimates (H_j') hold. Thus, we have global existence and smoothness for the sine-Gordon equation. We remark that in addition to leaving out some of the estimates, we have been treating B like differentiation and we have paid no attention to questions like domains of operators, etc. All of these technical details can be proven in similar ways as the details in Section 2, part A. We summarize.

<u>Theorem 11</u> If the initial data are real and in $C_0^\infty(R)$, then the sine-Gordon equation (22) has a unique global C^∞ solution if $m = 1,2,3,4$.

In Segal's original paper [31], he proves a more difficult result than Theorem 8. Namely, he shows that if J and its derivatives take the right domains of A^j into each other (this is implied by the estimates we assume) then the same conclusions hold. This is an interesting result but in practice the two results are the same since the only way one can prove that J takes the domains of powers of A into themselves is by having the estimates. W. von Wahl has pointed out that if one has global existence in escalated energy spaces then one can get the same conclusions as in the corollary to Theorem 9 just by assuming (H_j^L) and (H_j') for j large.

4. Finite propagation speed and continuous dependence on the data

Recovering classical smoothness from the abstract setting was relatively difficult in that it took more work and estimates. In contrast, finite propagation speed and continuous dependence on the data are easy.

Theorem 12 Let A be a self-adjoint operator on a Hilbert space $\mathcal{H}$ and J a non-linear mapping satisfying (H_o^L). Let $(-T,T)$ be the interval of existence of the solution $\phi(t)$ of (7) guaranteed by Corollary 1 of Theorem 1. Suppose that $\{P_t\}_{t\in(-T,T)}$ is a family of closed subspaces of $\mathcal{H}$ so that

$$e^{-iA(t_2-t_1)}:P_{t_1}\longrightarrow P_{t_2} \qquad \text{if} \quad T>t_2\geq t_1\geq o \tag{31}$$

$$e^{-iA(t_2-t_1)}:P_{t_1}\longrightarrow P_{t_2} \qquad \text{if} \quad -T<t_2\leq t_1\leq o \tag{32}$$

and

$$J:P_t\longrightarrow P_t \qquad \text{for all} \quad t\in(-T,T)$$

Then, if $\phi_o\in P_o$, we have $\phi(t)\in P_t$ for all $t\in(-T,T)$.

Proof We just use the same proof as for Corollary 1 of Theorem 1 except that we take for $\tilde{X}(T,\alpha,\phi_o)$ the set of continuous $\mathcal{H}$-valued functions $\psi(t)$ on $(-T,T)$ which satisfy $\psi(o)=\phi_o$, $\sup_{t\in(-T,T)}\|\psi(t)-e^{-itA}\phi_o\|\leq\alpha$ and $\psi(t)\in P_t$ for each $t\in(-T,T)$. $\tilde{X}(T,\alpha,\phi_o)$ is again a complete metric space and if we define

$$(S\psi)(t)=e^{-iAt}\phi_o+\int_o^t e^{-iA(t-s)}J(\psi(s))\,ds$$

then all the estimates are as in the Corollary to Theorem 1. We must just check that S takes $\tilde{X}(T,\alpha,\phi_o)$ into itself (as far as the P_t property is concerned). Suppose that $\psi(\cdot)\in\tilde{X}(T,\alpha,\phi_o)$. Then $\psi(s)\in P_s$ for each s and by the hypothesis on J, $J(\psi(s))\in P_s$ also. Now, suppose $t>o$. Then by (31), for each s satisfying $o\leq s\leq t$ we have

$$e^{-iA(t-s)}J(\psi(s))\in P_t$$

and as in the proof of Theorem 1, it is a continuous function of s. Therefore,

$$\int_0^t e^{-iA(t-s)} J(\psi(s))ds \in P_t$$

since P_t is closed. Since $e^{-iAt}\phi_0 \in P_t$ we conclude that $(S\psi)(t) \in P_t$ for $t \in [o,T)$ and a similar proof using (32) shows that $(S\psi)(t) \in P_t$ for $t \in (-T,o]$. Therefore S is a contraction on $\tilde{X}(T,\alpha,\phi_0) \subset X(T,\alpha,\phi_0)$ so its unique fixed point lies in $\tilde{X}(T,\alpha,\phi_0)$. Thus, $\phi(t) \in P_t$ for each $t \in (-T,T)$. ∎

<u>Example:</u> Let $\mathcal{H}$ be one of the Hilbert spaces discussed in Section 2. That is

$$\mathcal{H} = \mathcal{H}_{k/2} = D(B^{k+1}) + D(B^k)$$

where $B = \sqrt{-\Delta + m^2}$ on R^n and k is any non-negative integer. Let Σ be a compact set in R^n and define Q_t to be the set of $u \in L^2(R^n)$ so that the support of u is contained in the set $S(\Sigma,t) = \{x \in R^n |$ there exist $y \in \Sigma$ and $z \in R^n$ with $|z| \leq |t|$ and $x = y + z\}$ and let

$$P_t = (D(B^{k+1}) \cap Q_t) \oplus (D(B^k) \cap Q_t)$$

It is easy to check that $\{P_t\}$ is a family of closed subspaces. The conditions (31) and (32) are just the statement that the linear equation

$$u_{tt} - \Delta u + m^2 u = o$$

has propagation speed equal to one. This can be proven by integration by parts (for smooth solutions), by the explicit form of the solution (see [9],p.695) or by the Fourier transform and the Paley - Wiener theorem (see [27],p. 309). In any case this is a statement about the linear equation so we won't reproduce the proof here. Now, it is clear that all the non-linear terms $J(\cdot)$ which we have considered have the property

$$J : P_t \longrightarrow P_t$$

for each $t \in (-T,T)$. Thus, we conclude from Theorem 12, that if $\phi_0 \in P_0$ then $\phi(t) \in P_t$ for each t, i.e. the non-linear equation has propagation speed one. This is true wherever the solution exists, whether globally in t or only in a finite t interval. We summarize what we have proven:

Theorem 13 In all the examples discussed in Section 2 the solution propagates at speed one. In particular, for $C_o^\infty(R^n)$ data, the solution remains in $C_o^\infty(R^n)$.

Notice that this shows that the blow up of the L^2-norm in the non-global existence example in Section 2, part E, is caused by the function getting large locally, not because it fails to decay sufficiently fast at infinity.

* * *

Now, we treat the question of continuous dependence on the initial data. Whenever we have the hypothesis (H_o^L) of Theorem 1, then, according to the corollary, on each ball

$$\mathcal{B}(r) = \{\phi_o \in \mathcal{H} \,|\, ||\phi_o|| \leq r\}$$

we can at least solve the integral equation (7) (for $\phi_o \in \mathcal{B}(r)$) on some interval $(-T(r), T(r))$. For $t \in (-T(r), T(r))$ we define a non-linear mapping

$$M_t : \mathcal{B}(r) \longrightarrow \mathcal{H}$$

by

$$\phi(o) \xrightarrow{M_t} \mathcal{H}$$

Notice, that no M_t (except $M_o = I$) is necessarily defined on all of $\mathcal{H}$ since $T(r)$ may get smaller as $r \longrightarrow \infty$. We know however that where M_t exists it is strongly continuous as a function of t and by local uniqueness satisfies

$$M_t M_s = M_{t+s} \tag{33}$$

Now, continuous dependence on the data says that for each fixed t, M_t

is a continuous mapping (in the topology of $\mathcal{H}$) where it is defined. For general non-linear mappings boundedness is not enough to guarantee continuity but in conjunction with (H_0^L) it is. Since we are for the moment just using (H_0^L), by "solution exists" we mean a continuous solution of

$$\phi(t) = e^{-itA}\phi_0 + \int_0^t e^{-iA(t-s)} J(\phi(s))ds \qquad (7)$$

where we just assume $\phi_0 \in \mathcal{H}$.

Theorem 14 Let A be a self-adjoint operator on a Hilbert space $\mathcal{H}$ and let J be a non-linear mapping on $\mathcal{H}$ satisfying hypothesis (H_0^L) of Theorem 1. Let D be a (not necessarily closed) set in $\mathcal{H}$, and $(-T,T)$ a finite interval so that for $\phi_0 \in D$, $M_t\phi_0$ exists and

$$||M_t\phi_0|| \leq K(T,D)$$

for all $t \in (-T,T)$ where K depends on T and D but not on ϕ_0. Then for each t M_t is uniformly continuous on D.

Proof The proof is almost trivial. Let $\phi_0^{(1)}, \phi_0^{(2)} \in D$ and define $\phi_i(t) = M_t\phi_0^{(i)}$. Then, since each $\phi_i(t)$ satisfies the integral equation (7) on $(-T,T)$ with corresponding initial data, we have,

$$||\phi_1(t) - \phi_2(t)|| \leq ||\phi_0^{(1)} - \phi_0^{(2)}|| + \int_0^t ||J(\phi_1(s)) - J(\phi_2(s))||ds$$

$$\leq ||\phi_0^1 + \phi_0^2|| + \int_0^t C(||\phi_1(s)||,||\phi_2(s)||)\ ||\phi_1(s) - \phi_2(s)||ds$$

$$\leq ||\phi_0^1 - \phi_0^2|| + C(K,K)\int_0^t ||\phi_1(s) - \phi_2(s)||ds$$

so

$$(34) \qquad ||\phi_1(t) - \phi_2(t)|| \leq ||\phi_0^{(1)} - \phi_0^{(2)}||\ e^{tC(K,K)} \qquad \blacksquare$$

Corollary 1 Assume hypotheses (H_0^L). Let $(-T_r,T_r)$ be the finite interval of existence of solutions of (7) constructed by the method of the Corollary to Theorem 1 for all ϕ_0 in $\mathcal{B}(r) = \{\phi_0 \mid ||\phi_0|| \leq r\}$. Then for each $t \in (-T_r,T_r)$, M_t is uniformly continuous on $\mathcal{B}(r)$.

Proof From the proof of the Corollary to Theorem 1 it follows that $||M_t\phi_0|| \le K + \alpha$ for all $\phi_0 \in \mathcal{B}(r)$ so the hypotheses of Theorem 14 are satisfied with $D = \mathcal{B}(r)$.

Corollary 2 Suppose the hypotheses of Theorem 14. Then for all ϕ_0 in the closure of D the solution of (7) exists on $(-T,T)$; these solutions are uniformly bounded on $(-T,T)$ with the same bound and M_t is uniformly continuous on $\bar{D}$ for each $t \in (-T,T)$.

Proof Let $\phi_0 \in \bar{D}$ and let $\phi_0^{(n)}$ be a sequence of vectors in D which converge to ϕ_0. Let $\phi_n(t)$ be the corresponding solutions on $(-T,T)$. Then by the estimate (34) the $\phi_n(t)$ converge uniformly to a continuous $\mathcal{H}$-valued function $\phi(t)$ on $(-T,T)$. Since $J(\phi_n(t))$ converges uniformly to $J(\phi(t))$ (by the hypothesis (H_0^L)), and each $\phi_n(t)$ satisfies

$$\phi_n(t) = e^{-itA}\phi_0^{(n)} + \int_0^t e^{-iA(t-s)}J(\phi_n(s))ds$$

we can take the limit as $n \longrightarrow \infty$ and conclude that $\phi(t)$ satisfies (7) for $t \in (-T,T)$. Since the $\phi_n(t)$ are uniformly bounded, $\phi(t)$ is bounded with the same bound on $(-T,T)$. Now, applying Theorem 14 to $\bar{D}$ we get that M_t is uniformly continuous on $\bar{D}$ for each $t \in (-T,T)$. ∎

Examples: First, it is clear that the first corollary implies continuous dependence for short times for all the examples considered in Section 2 including those where the solution blows up after a finite amount of time.

The application to the sine-Gordon equation is also easy. In that case

$$||\phi(t)|| \le ||\phi_0||e^t$$

for all $\phi_0 \in \mathcal{H}$. Thus we just choose D to be the ball $\mathcal{B}(r) = \{\phi_0 \mid ||\phi_0|| \le r\}$ and apply Theorem 14 directly to conclude that on any finite time interval $(-T,T)$, M_t is a uniformly equicontinuous family of mappings.

The application to the case $J(\phi) = \langle o,-u^3\rangle$, $n = 3$, is a little more subtle and uses Corollary 2. Notice that in this case we proved global existence of solutions of

$$\phi'(t) = -iA\phi(t) + J(\phi(t)) \quad , \quad \phi(o) = \phi_0$$

if $\phi_0 \in D(A)$. In case ϕ_0 is just in $\mathcal{H}$ then we don't expect the differential equation to hold, but the integral equation (7) has a local solution because (H_0^L) holds. However, we don't know yet whether (29) has a global solution since we have no direct way of getting an apriori bound on $||\phi(t)||$. The reason is that we can't conclude that the energy $E(t)$ is conserved since we can't differentiate $E(t)$. Nevertheless we can handle this difficulty by using Corollary 2. Let $D = \mathcal{B}(r) \cap D(A)$. Then for $\phi_0 \in D$, energy is conserved. Using the estimate $(u(x,o) = u_o)$

$$\begin{aligned}\int_{R^3} |u_0|^4 dx &\leq ||u_0||_2 \; ||u_0||_6^3 \\ &\leq ||u_0||_2 K \; ||Bu_0||_2^3 \\ &\leq K \; ||Bu_0||_2^4\end{aligned}$$

and conservation of energy, we have (for all $\phi_0 \in \mathcal{B}(r)$)

$$\begin{aligned}||\phi(t)||^2 &\leq ||\phi(t)||^2 + \frac{1}{2}\int_{R^3} |u(t)|^4 dx \\ &= 2E(t) = 2E(o) \\ &= ||\phi(o)||^2 + \frac{1}{2}\int |u_0|^4 dx \\ &\leq ||\phi(o)||^2 + K||\phi(o)||^4 \\ &\leq r + Kr^2 \qquad (35)\end{aligned}$$

Thus, for every finite interval $(-T,T)$ the hypotheses of Corollary 2 are satisfied. Therefore, for every $\phi_0 \in \bar{D} = \mathcal{B}(r)$, the integral equation has a solution on the whole interval which has the same bound. Since this is true for all $(-T,T)$, the solution is global and since the estimate (35) is independent of T the family $\{M_t\}$ which is now defined on all of $\mathcal{H}$ for all t is uniformly equicontinuous.

This same proof works for all the cases where we got global existence of strong solutions. Note that in the case $m = o$, considered in part B of Section 2, we can conclude that M_t is uniformly continuous for each t, but only equicontinuous on finite $(-T,T)$ intervals because the estimate corresponding to (35) depends on T.

Finally, we make two remarks. By using the escalated energy spaces or by using the trick of Theorem 14 applied successively for $j = 1,\ldots,m$, (in the cases where $(H_j^{!})$, (H_j^{L}) hold), one can conclude that

$$\sum_{j=1}^{m} ||A^j(\phi(t) - \psi(t))|| \leq C(t) \sum_{j=1}^{m} ||A^j(\phi(o) - \psi(o))||$$

if $\phi_o, \psi_o \in D(A^m)$. So, if a certain number of derivatives of the initial data are close enough then the same number of derivatives of the solution will be close. Secondly, since we can explicity estimate the constants involved in all the problems we have discussed, we have an explicit modulus of continuity (which may depend on t). We summarize all of this discussion in the following (somewhat vague) theorem.

Theorem 15 In all the examples in Section 2, the solutions depend continuously on initial data.

5. Weak Solutions

In this section we will show that the equation

$$u_{tt} - \Delta u + m^2 u = - u^p, \qquad x \in R^n \tag{36}$$

$$u(x,o) = f(x)$$

$$u_t(x,o) = g(x)$$

has global weak solutions for all odd $p > o$ and all n. There are several proofs (Segal [30], Lions [19], Strauss [36]) but all use basically the same idea. One regularizes the right hand side in such a way that one obtains global existence for the solution $u_n(t)$ of the regularized problem. One then lets the regularized right hand side approach $-u^p$ and extracts a convergent subsequence of the $\{u_n(t)\}$ by a compactness argument based on uniform energy estimates. By a limiting argument (again using the energy inequality) one then shows that $u(t)$ satisfies (36) in a weak sense. We will follow the argument of Strauss which applies most naturally if we restrict our attention to real-valued functions. Therefore, <u>for this section only</u>:

$$A = - \begin{pmatrix} o & I \\ -B^2 & o \end{pmatrix}$$

$$B = \sqrt{- \Delta + m^2}$$

$$\mathcal{H} = D(B) \oplus L^2(R^n)$$

where $D(B)$ and $L^2(R^n)$ are spaces of <u>real-valued functions</u>. A is skew-adjoint on $D(A) = D(B^2) \oplus D(B)$ and of course generates on $\mathcal{H}$ the group $W(t) = e^{-tA}$ given in the introduction. All the considerations of the past sections go over to this case so we use them without comment.

We will take the initial data f and g to be in $C_o^\infty(R^n)$, it will be clear from the argument that we could take a much more general class. Define $F_n(x)$ to be the continuous function on R satisfying:

$$F_n(x) = \begin{cases} x^p & |x| \le n \\ \text{linear} & n \le |x| \le n+1 \\ 0 & |x| \ge n+1 \end{cases}$$

and let

$$G_n(x) = \int_0^x F_n(y)\,dy$$

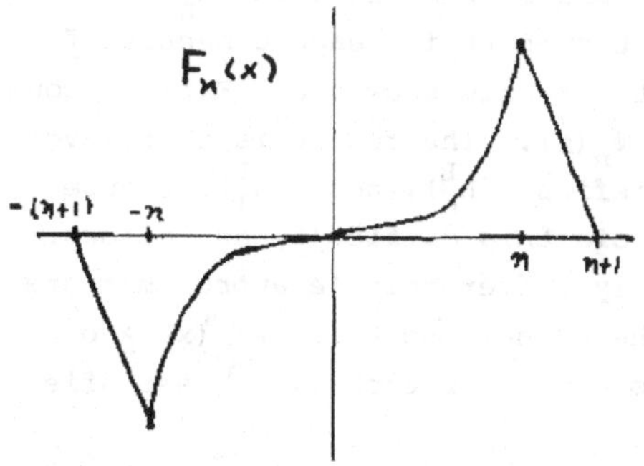

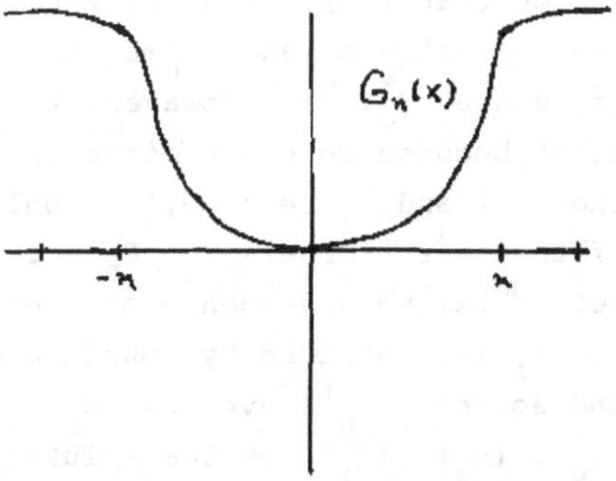

Finally, define

$$J_n(\phi) = J_n(<u,v>) = <0,-F_n(u)>$$

Then, since $F_n(x)$ is a Lipschitz function, it is easy to see that J_n satisfies (H_0^L). Thus, by Theorem 1 there is a local solution $\phi_n(t) = <u_n(t),\, u_n'(t)>$ which satisfies

$$\text{(37)} \qquad \phi_n(t) = e^{-At}\phi_0 + \int_0^t e^{-A(t-s)} J_n(\phi_n(s))\,ds$$

$$\phi_n(0) = <f,g>$$

The function $u_n(t)$ satisfies

$$\text{(38)} \qquad u_n(t) = \cos(Bt)f + \frac{\sin(Bt)}{B}\,g + \int_0^t \left[B^{-1}\sin B(t-s)\right] F_n(u(s))\,ds$$

and formally satisfies:

$$u_{tt} - \Delta u + m^2 u = - F_n(u) \tag{39}$$

Thus, $u_n(t)$ should have the conserved energy

$$E_n(t) = \frac{1}{2}||Bu_n(t)||_2^2 + \frac{1}{2}||u_n'(t)||_2^2 + \int_{R^n} G_n(u_n(t))dx \tag{40}$$

Notice that (40) certainly makes sense and is finite since G_n is bounded, $G_n(o) = o$, and $u_n(t)$ has compact support for each t because f and g are $C_o^\infty(R^n)$. However, we cannot directly show that E(t) is constant because we can't differentiate $\phi_n(t)$. The reason is that, even though f and g are nice, J_n only satisfies (H_o^L), not (H_1^L) because of the sharp corners in $F_n(x)$. We avoid this difficulty as follows. Let $F_n^{(m)}(x)$ be a sequence of continuously differentiable approximations to $F_n(x)$, obtained by rounding off the corners so that $xF_n^{(m)}(x) \geq o$ and so that $F_n^{(m)}$ converges to F_n uniformly. For each m, $J_n^{(m)}$ satisfies (H_o^L), (H_1'), (H_1^L) so the solution of

$$\phi_n^{(m)}(t) = e^{-At}\phi_o + \int_o^t e^{-A(t-s)} J_n^{(m)}(\phi_n^{(m)}(s))ds \tag{41}$$

is strongly continuously differentiable. Thus we can differentiate

$$E_n^{(m)}(t) = \frac{1}{2}||Bu_n^{(m)}(t)||_2^2 + \frac{1}{2}||\frac{d}{dt}u_n^{(m)}(t)||_2^2 + \int_{R^n} G_n^{(m)}(u_n^{(m)}(t))dx \tag{42}$$

and prove that it is conserved. From this it follows that $||\phi_n^{(m)}(t)||$ is apriori bounded (since $G_n^{(m)}(x) \geq o$) so the solution of (41) is global by Theorem 2. Now let $[-T,T]$ be any subinterval of the interval on which a continuous solution of (37) exists. Then both $||\phi_n(t)||$ and $||\phi_n^{(m)}(t)||$ are uniformly bounded on $[-T,T]$ so by our usual iteration trick (see for example, Theorem 14) we can prove that $||\phi_n^{(m)}(t) - \phi_n(t)||$ $\longrightarrow o$ uniformly on $[-T,T]$. From this it easily follows that $E_n^{(m)}(t) \longrightarrow E_n(t)$ for $t \in [-T,T]$ and thus, since each $E_n^{(m)}(t)$ is constant we conclude that $E_n(t)$ is constant. Finally, this implies as usual that $||\phi_n(t)||$ is apriori bounded (since $G_n(x) \geq o$) and so the solution $\phi_n(t)$ of (37) is global and $E_n(t)$ is constant for all t.

Now we come to the main part of the argument. Since f is nice,

$$E_n(o) = \frac{1}{2}||Bf||_2^2 + \frac{1}{2}||g||_2^2 + \int G_n(f)dx$$

converges as $n \longrightarrow \infty$ to a number

$$E(o) \equiv \frac{1}{2}||Bf||_2^2 + \frac{1}{2}||g||_2^2 + \int_{R^n} (f(x))^{p+1}dx$$

Therefore the numbers $\{E_n(o)\}$ are uniformly bounded. But since $E_n(t)$ is constant in t for each n, this means that there is a constant C so that

(43) $\qquad E_n(t) \leq C \qquad$ for all t and n.

Let $S(r)$ be the ball in R^n of radius r and choose r_0 so that the supports of f and g lie in $S(r_0)$. Let $[-T,T]$ be a given finite interval. Then by (43) and (40)

$$||u_n'(t)||_2 \leq \sqrt{2C}$$

so $u_n(t)$ are a uniformly equicontinuous family of functions on $[-T,T]$ with values in $L^2(S(r_0 + T))$. But (again by (40) and (43)) the values lie in

$$\{v \in L^2(S(r_0 + T)) \mid ||v||_2 \leq \sqrt{2C/m}, ||Bv||_2 \leq \sqrt{2C}\}$$

Since this set is compact in $L^2(S(r_0 + T))$, the Ascoli-Arzela theorem (see [45], p. 155) tells us that $\{u_n(t)\}$ has a convergent subsequence (which we also call $u_n(t)$) so that $u_n(t)$ converges uniformly to a continuous $L^2(S(r_0 + T))$-valued function $u(t)$ on $[-T,T]$. By the usual diagonalization trick (for larger and larger T) we can assume that we have a subsequence so that this statement holds for each T.

Let v be in $C_0^\infty(R^n)$ (again, this is stronger than necessary). Then by (38),

(44) $$(u_n(t),v) = (\cos(Bt)f,v) + (B^{-1}\sin(Bt)g,v)$$
$$+ \int_0^t (-B^{-1}\sin B(t-s)]\, F_n(u_n(s)),v)ds$$

$$= (\cos(Bt)f,v) + (B^{-1}\sin(Bt)g,v) + \int_0^t (- F_n(u_n(s)), B^{-1}\sin B(t-s)v)ds$$

Suppose that we can show that

$$(45) \qquad F_n(u_n(x,t)) \xrightarrow{L^1(R^n \times [-T,T])} u(x,t)^p$$

then since $B^{-1}\sin(B(t-s))v$ is a C^∞ function of all its variables we can take the limit in (44) to conclude that:

$$(46) \quad (u(t),v) = (\cos(Bt)f,v) + (B^{-1}\sin(Bt)g,v) + \int_0^t (-u(s)^p, [B^{-1}\sin B(t-s)]v)dx$$

Since f and g are nice and the integrand on the right is in L^1 , the left side is absolutely continuous and

$$\frac{d}{dt}(u(t),v) = -(B\sin(Bt)f,v) + (\cos(Bt)g,v) + \int_0^t (-u(s)^p, [\cos B(t-s)]v)ds$$

Again, the right hand side is absolutely continuous, so

$$\frac{d^2}{dt^2}(u(t),v) = (\cos(Bt)f, -B^2v) + (B^{-1}\sin(Bt)g, -B^2v)$$

$$+ (- u(t)^p,v) + \int_0^t (- u(s)^p,[B^{-1}\sin B(t-s)](- B^2v))ds$$

$$= (u(t), -B^2v) + (- u(t)^p,v)$$

for almost all t. Thus,

$$\frac{d^2}{dt^2}(u(t),v) - (u(t),\Delta v) + m^2(u(t),v) = (- u(t)^p,v)$$

$$(u(o),v) = (f,v)$$

$$\frac{d}{dt}(u(t),v)\Big|_{t=o} = (g,v)$$

so $u(t,x)$ is a weak global solution of

$$u_{tt} - \Delta u + m^2 u = - u^p$$

It remains to prove (45) by a real variables argument and clever use of (43) (again!). Let $[-T,T]$ be a fixed finite time interval. Since $u_n(t) \xrightarrow{L^2} u(t)$ uniformly on $[-T,T]$ we have

$$u_n \xrightarrow{L^2([-T,T] \times R^n)} u$$

so we can choose a subsequence (again denoted by u_n) so that $u_n \longrightarrow u$ pointwise a.e. in $[-T,T] \times R^n$. It follows immediately from properties of the F_n that

(46) $$F_n(u_n(t,x)) \longrightarrow u(x,t)^p$$

pointwise a.e. in $[-T,T] \times R^n$. Since $|F_n(x)| \le 1 + G(x)$ we have

$$\int_{-T}^{T} \int_{R^n} |F_n(u_n(x,t))| \, dxdt \le 2T \operatorname{Vol}[S(r_o + T)] + \int_{-T}^{T} \int_{R^n} G_n(u_n(x,t)) dxdt$$

$$\le 2T \operatorname{Vol}[S(r_o + T)] + 2TC$$

where we have used the finite propagation speed and (43). Thus, by Fatou's lemma,

$$\int_{-T}^{T} \int_{R^n} |u(x,t)|^p \, dxdt \le \underline{\lim} \int_{-T}^{T} \int_{R^n} |F_n(x,t)| \, dxdt < \infty$$

so $|u|^p \in L^1([-T,T] \times R^n)$. Now, by the finite propagation speed, $F_n(u_n)$ and u^p have support in $S(r_o + T)$ for $|t| \le T$ so (46) implies by Egorov 's theorem that $F_n(u_n(x,t))$ converges to $u(x,t)^p$ uniformly except on a set of arbitrarily small measure δ in $[-T,T] \times S(r_o + T)$. Therefore to conclude (45) we need only show that, given ε, we can choose δ so that

$$\iint_M |F_n(u_n(x,t))|\, dxdt < \varepsilon$$

whenever the measure of M in $[-T,T] \times S(r_o + T)$ is less than δ. Now, $F_n(x)$ is only large if $|x|$ is large so given $2TC/\varepsilon$ we can find a constant K so that $|F_n(x)| \geq K$ implies $|x| \geq 2TC/\varepsilon$. Choose δ so that $\delta K \leq \varepsilon/2$ and for any M write $M = M_n' \cup M_n''$ where

$$M_n' \equiv \{ \langle x,t \rangle \mid |F_n(u_n(x,t))| \geq K \}$$

and M_n'' is its complement in M. Then

$$\begin{aligned}
\iint_M |F_n(u_n)|\, dxdt &= \iint_{M_n'} |F_n(u_n)|\, dxdt + \iint_{M_n''} |F_n(u_n)|\, dxdt \\
&\leq \frac{\varepsilon}{4TC} \iint |u_n(x,t)|\,|F_n(u_n)|\, dxdt + K\delta \\
&\leq \frac{\varepsilon}{4TC} \iint G_n(u_n(x,t))\, dxdt + \varepsilon/2 \\
&\leq \varepsilon \qquad \text{(by (43))}
\end{aligned}$$

In the next to last step we have used the fact that $|x|\,|F_n(x)| \leq G_n(x)$ for each n and all x. This holds because $F_n(x)$ is monotone decreasing to the right of n which is why we needed the sharp corner in the definition of F_n.

As we remarked before, this proof follows the outline in Strauss [36]. Strauss actually proves the more general result that global weak existence holds for

$$u_{tt} - \Delta u + m^2 u = F(u)$$

as long as $F(x)$ is a continuous real-valued function satisfying $xF(x) \leq 0$. The general ideas come from Segal's paper [30] but Segal chooses to regularize u^p by $j*(u*j)^p$ where j is an approximate identity. This makes it easy to handle the complex-valued case but makes the proof of the convergence of (44) as $n \longrightarrow \infty$ more difficult. We remark that in these proofs one loses uniqueness because of the compactness argument.

6. Discussion

Before going on to scattering theory it is worthwhile to discuss what we have presented and to point out worthwhile research problems. The best aspects of the abstract theory are that it is simple (the abstract proofs were quite easy and in general followed ideas from ordinary differential equations) and that it is quite general in that the hypotheses involve A and J directly and not special properties of the group e^{-itA} which is sometimes difficult to compute. As we will see, to construct an abstract scattering theory one needs much more information on e^{-itA}.

We have chosen examples which illustrate easily the applications of the abstract theory and its limitations without trying to give general conditions on functions $f(x,u,u_t)$ so that

$$u_{tt} - \Delta u + m^2 u = f(x,u,u_t)$$

has local or global solutions. Such general conditions may be found in Strauss [34] for the case $m = 0$ and in Chadam [3] for the case $m > 0$. There are a wide range of other equations to which this abstract theory can be applied. Specific examples can be found in most of the papers listed in the index. See in particular the references in the paper by Scott, Chu, and McLaughin [29]. There are many equations of physical interest for which the details of the existence theory have not been worked out and to which these methods can be applied. Essentially there are problems in application of the theory as it now stands.

There are two problems one specific and one general which are difficult but whose solution would lead, in my opinion, to great progress in non-linear partial differential equations. The specific problem is of course our old friend:

$$u_{tt} - \Delta u + m^2 u = - u^p \qquad x \in R^3$$

for high odd p. We know that strong smooth local solutions exist if the data are nice and we know that global weak solutions exist. Some new idea is required to prove that the solution does not keep losing derivatives. If one could solve this problem, then I'm sure one could solve a host of other problems where there is no strong existence theory because of the lack of Sobolev estimates.

The second interesting question is to investigate problems where either global existence is false or where it is unknown and to try to prove global existence for certain subclasses of initial data. As one example of this we will prove in Section 10 that

$$u_{tt} - \Delta u + m^2 u = \lambda u^p$$

has a global strong solution for high p in three dimensions if the initial data are small enough. A more interesting example is provided by the work of Chadam and Glassey on the Yukawa coupled Dirac and Klein-Gordon equation in three dimensions:

$$(-i\gamma^u \partial_u + M)\psi = g\phi\psi \tag{48a}$$

$$u_{tt} - \Delta u + m^2 u = g\overline{\psi}\gamma^o\psi \tag{48b}$$

where $M > 0, \partial = \langle \partial_t, \partial_{x_1}, \partial_{x_2}, \partial_{x_3} \rangle$, the γ's are certain four by four matrices, and

$$\psi = \langle \psi_o(x,t), \psi_1(x,t), \psi_2(x,t), \psi_3(x,t) \rangle$$

$\overline{\psi}$ always denotes the complex conjugate. Because of special properties of the γ's, there is a conserved energy

$$E(t) = \int_{R^3} \overline{\psi}\gamma_o \left(i\sum_{k=1}^{3} \gamma^k \partial_k + M\right)\psi + \int_{R^3} (\nabla u)^2 + m^2u^2 + u_t^2 - g\int_{R^3} \phi\overline{\psi}\gamma_o \psi$$

but it is not positive definite. Furthermore, because of the first degree of the time derivative in (48a) one must use an escalated energy space to prove local existence. Therefore, it is not clear how much good the conserved energy would be even if it were positive. Thus, these equations have both the difficulty of (47) for high odd p and in addition the fact that the energy can be negative. Work on these equations goes back to Gross [13] who investigated the coupled Maxwell-Dirac equations where similar problems arise. Gross proved that local solutions exist. Chadam [4] then proved that solutions exist in an arbitrary region of space time if g or the initial data are small enough. Then, in [5], Chadam showed that the corresponding equations in one dimension have global solutions for arbitrary initial data (this is the proof that we gave in Section 2, part F).

In three dimensions the question of global existence for (48) is open, but the following partial result of Chadam and Glassey [7] is very suggestive. They show that whenever a solution exists then

$$\int_{R^3} \{|\psi_1 - \overline{\psi}_4|^2 + |\psi_2 + \overline{\psi}_3|^2\}dx \tag{49}$$

is conserved. Thus, if initially, $\psi_1(x,o) = \overline{\psi_4(x,o)}$ and $\psi_2(x,o) = -\overline{\psi_3(x,o)}$ then the same will be true for all time. From this it follows that (in the standard representation of the γ's)

$$\overline{\psi}\gamma_0\psi = |\psi_1|^2 + |\psi_2|^2 - |\psi_3|^2 - |\psi_4|^2 = o$$

for all time. Thus, for such initial data ϕ satisfies the free Klein-Gordon equation and one can use this to show global existence for (48). So, for a certain subspace of initial data global solutions exist.

These results suggest that in general for systems without positive energy (and where global existence does not or is not known to hold for all data) there may be whole subspaces of initial data (determined by some algebraic conditions) for which global existence does hold. Such proofs will require the discovery and clever use of new conserved quantities. It is not clear, of course, what the physical significance of these subspaces will be, but this whole question seems to me to be very interesting and worthwhile.

Chapter 2 Scattering Theory

7. Formulation of Scattering Problems

In this lecture we provide background and motivation for developments in the rest of the chapter. Let us begin by looking at the equation

$$u_{tt} - \Delta u + m^2 u = -\lambda u^p \qquad (50)$$

whose existence theory for some $p > 1$ we treated in Chapter 1. To develope a scattering theory for (50) one would show that for large positive and negative times the solutions of (50) look more and more like solutions of the corresponding free equation:

$$u_{tt} - \Delta u + m^2 u = o \qquad (51)$$

It is not clear a priori why this should be true since the non-linear term $-\lambda u^p$ not only creates an interaction throughout all of space but is clearly very large if u is large. On the other hand, the solutions of the free equation (51) with nice initial data decay (in the sup norm) like $t^{-n/2}$ in n dimensions. This raises of the possibility that the same decay will hold for solutions of (50) in which case a scattering theory might be possible since $-\lambda u^p$ is very small when u is small. Thus, we expect the existence of a scattering theory will depend on a delicate interplay between the rate of decay, the degree of the non-linear term, and the sign of the non-linear term since we know from Section 2, part E that some solutions blow up in finite time if $\lambda < o$ or if p is even.

As in Chapter 1 our technique is to reformulate the scattering theory problem as an abstract Hilbert space problem, prove abstract results, and then return to applications. So let A be a self-adjoint operator on a Hilbert space $\mathcal{H}$ and J a non-linear mapping from $\mathcal{H}$ to itself so that

$$\phi(t) = e^{-itA}\phi_o + \int_o^t e^{-iA(t-s)} J(\phi(s))ds \qquad (52)$$

has global solutions at least for some initial data ϕ_o. As indicated in Chapter 1, if J satisfies certain additional conditions and $\phi_o \in D(A)$ then $\phi(t)$ is differentiable and satisfies

$$\phi'(t) = -iA\phi(t) + J(\phi(t))$$

However, in this chapter we concentrate on the asymptotic behavior and do not worry about smoothness, so we will just work with the integral equation (52). What we would like to prove is that there is a nice set of initial data $\Sigma_{scat} \subset \mathcal{H}$ with the following properties:

(a) For each $\phi_- \in \Sigma_{scat}$ there is a solution $\phi(t)$ of (52) so that

$$\phi(t) - e^{-itA}\phi_- \longrightarrow 0 \quad \text{as} \quad t \longrightarrow -\infty.$$

That is, as $t \longrightarrow -\infty$ $\phi(t)$ looks more and more like the solution $e^{-itA}\phi_-$ of the free equation with initial data ϕ_-.

(b) For each $\phi_+ \in \Sigma_{scat}$ there is a solution $\tilde{\phi}(t)$ of (52) so that

$$\tilde{\phi}(t) - e^{-itA}\phi_+ \longrightarrow 0 \quad \text{as} \quad t \longrightarrow +\infty.$$

That is, $\tilde{\phi}(t)$ looks more and more like the free solution $e^{-itA}\phi_+$ with initial data ϕ_+ as $t \longrightarrow +\infty$.

If (a) and (b) hold we define the <u>wave operators</u> Ω_+ and Ω_- by

$$\Omega_- : \phi_- \longrightarrow \phi(0)$$

$$\Omega_+ : \phi_+ \longrightarrow \tilde{\phi}(0)$$

(c) Further we would like to prove that

$$\text{Range } \Omega_+ = \text{Range } \Omega_-$$

(this is called the problem of <u>asymptotic completeness</u>) and also that Ω_+ and Ω_- are one to one. Then we can define the scattering operator

$$S : \Sigma_{scat} \longrightarrow \Sigma_{scat}$$

by

$$S = (\Omega_+)^{-1}\Omega_-$$

(d) Finally, one would like to prove some properties of S. For example, that S is a continuous mapping of Σ_{scat} onto itself if Σ_{scat} has an appropriate topology.

Basically, our approach to the problem is to try to solve the initial value problem for (52) with Cauchy data ϕ_- given at $t = -\infty$. That is, to show that the equation

$$\phi(t) = e^{-iAt}\phi_- + \int_{-\infty}^{t} e^{-iA(t-s)}J(\phi(s))ds \tag{53}$$

has a global solution, that $\phi(t) - e^{-iAt}\phi_- \longrightarrow o$ as $t \longrightarrow -\infty$, and that there exists a ϕ_+ so that $\phi(t) - e^{-iAt}\phi_+ \longrightarrow o$ as $t \longrightarrow +\infty$.

As we mentioned at the beginning, carrying through this complete program for highly non-linear equations with interactions throughout space is a very difficult problem. Complete results are known only in a few cases. However, two parts of the theory are easier. First if the initial data ϕ_- at $t = -\infty$ is small (or if the coupling constant is small) then one can prove global existence and for (53) and develop a scattering theory quite easily if the solutions $e^{-iAt}\phi_-$ of the free equation decay sufficiently rapidly and J has high enough degree. Essentially the reason in that the non-linear term never gets large enough to dominate the equation. We develop this scattering theory for small data in Section 8.

The second part of the problem that is fairly easy is to show the existence of the wave operators Ω_+ and Ω_-. What is required is that (52) have global solutions, that $e^{-itA}\phi_-$ decays sufficiently rapidly, and that J have high enough degree. The proof which uses many of the ideas of Section 8 is given in Section 10.

In Section 11 we give applications of the results in Sections 8,9, and 10 What remains in most cases is the problem of asymptotic completeness. This is a really difficult problem because (as we discuss in Section 12) its solution requires a priori decay estimates on solutions of the non-linear equations. Such estimates are known in only a few cases. However, <u>if</u> one has such an apriori estimate, then one can carry through the entire theory if J satisfies the right properties. This we do in Section 12.

It is worth mentioning before we start, two technical difficulties which do not occur in quantum mechanical scattering. First, since the wave and scattering operators are non-linear, it is not sufficient to prove that they are bounded on a dense set in order to conclude that they are continuous. Secondly, it is natural to take as scattering states, $\sum_{scat}$, the set of ϕ in $\mathcal{H}$ so that $e^{-itA}\phi$ decays appropriately. Unfortunately, in most applications, only sufficient but not necessary condition are known for such decay. Thus if we want in addition that $\sum_{scat}$ be closed (for example, to be a Banach space) then the norm of a ϕ in $\sum_{scat}$ must involve explicitly the large time behavior of $e^{-itA}\phi$. In what follows these two difficulties will cause quite a bit of technical pain.

Finally, as you will see, we will need quite strong hypotheses on J in order to develope as abstract scattering theory, many more hypotheses than we needed for the existence theory. This is natural since the existence of a scattering theory reflects a very close relationship between the solutions of the free and interacting equations. Thus, it requires quite strong hypotheses on J. We remark however, that the following situation sometimes occurs. In a particular application, J may not fulfill all the hypotheses of the appropriate theorem below. Nevertheless, one can often prove the conclusion of the theorem by using special properties of the particular equations and the idea of the general theorem. We will see an example of this in Section 11c.

8. Scattering for small data

In this lecture we show how to construct the scattering operator directly if the the data at $-\infty$ are sufficiently "small". The basic idea is that if the data is small and the non-linear term has high enough degree then the non-linear term should always remain small in comparison to the linear terms. From this one can get global existence and a scattering theory. The important point here is that we will use no energy inequalities (or apriori boundedness) at all. Thus these theorems provide global existence for small data in some cases where global existence fails for large data; for example, for

$$u_{tt} - \Delta u + m^2 u = u^p$$

for large p. This is consistent with the non-existence example in part E of Section 2 since we had to make the data "large" in order to get the solution to blow up in finite time.

We remark that if there is a small coupling constant in front of J, that is, if we are dealing with the integral equation that corresponds to

$$\phi'(t) = -iA\phi(t) + gJ(\phi(t)) \tag{54}$$

then the theorems in this section imply global existence for any (not necessarily small) initial data if g is small enough. However, the larger the data the smaller g must be chosen so we can't handle (by the methods of this section) all data for any one fixed g.

So, let A be a self-adjoint operator on $\mathcal{H}$. Let $\|\ \|_a$ and $\|\ \|_b$ be two auxiliary "norms" on $\mathcal{H}$: $\|\cdot\|_a$ satisfies all the properties of a norm except that $\|\phi\|_a = 0$ need not imply that $\phi = 0$; $\|\cdot\|_b$ satisfies all the properties of a norm except that it may take the value $+\infty$. We assume that A, J, and the norms $\|\cdot\|_a$, $\|\cdot\|_b$ satisfy the following hypotheses:

(i) There is a $c > 0$ so that

$$\|\phi\|_a \leq c\|\phi\| \qquad \text{for all} \quad \phi \in \mathcal{H} \tag{55}$$

(ii) There are constants $c_1 > 0$, $d > 0$ so that for $\phi \in \mathcal{H}$

$$||e^{-iAt}\phi||_a \leq c_1 t^{-d}||\phi||_b \quad \text{if} \quad |t| \geq 1 \tag{56}$$

(iii) There exist $\beta > o$, $\delta > o$, and $q \geq 1$ with $dq > 1$, so that

$$||J(\phi_1) - J(\phi_2)|| \leq \beta(||\phi_1||_a + ||\phi_1||_a)^q||\phi_1 - \phi_2|| \tag{57}$$

$$||J(\phi_1) - J(\phi_2)||_b \leq \beta\{(||\phi_1||_a + ||\phi_2||_a)^{q-1}||\phi_1 - \phi_2||_a + (||\phi_1||_a + ||\phi_2||_a)^q||\phi_1 - \phi_2||\} \tag{58}$$

for all $\phi_1, \phi_2 \in \mathcal{H}$ satisfying $||\phi_i|| \leq \delta$. In the case $q = 1$ we assume that β can be chosen arbitrarily small if δ is chosen small.

We can now define the scattering states and the scattering norm. First, for an $\mathcal{H}$-valued function $\psi(t)$ on R we define

$$|||\psi(\cdot)|||_{[N_1,N_2]} \equiv \sup_{N_1 \leq t \leq N_2} ||\psi(t)|| + \sup_{N_1 \leq t \leq N_2} (1+|t|)^d||\psi(t)||_a$$

In the case where $N_1 = -\infty$, $N_2 = +\infty$ we will denote the norm simply by $|||\cdot|||$. Now we define

$$\Sigma_{scat} \equiv \{\phi \in \mathcal{H} \mid |||e^{-itA}\phi||| < \infty\}$$

and

$$||\phi||_{scat} \equiv |||e^{-itA}\phi|||$$

That is, the scattering states are just those vectors in $\mathcal{H}$ which decay nicely under the free propagation. Notice that if $||\phi||_b < \infty$, then

$$||e^{-itA}\phi||_a \leq c_2(1+|t|)^{-d}(||\phi|| + ||\phi||_b) \quad \text{for all } t \tag{59}$$

where $c_2 = \max\{2^d c_1, 2^d c\}$, so $\phi \in \Sigma_{scat}$ and

$$||\phi||_{scat} \leq (1+c_2)||\phi|| + c_2||\phi||_b$$

We can now state our first main theorem:

Theorem 16 (global existence for small data)

Let A be a self-adjoint operator on a Hilbert space $\mathcal{H}$ and J a non-linear mapping of $\mathcal{H}$ into itself. Suppose that there exist norms $||\cdot||_a$, $||\cdot||_b$ so that hypotheses (i), (ii), and (iii) hold. Then there is an $\eta_o > o$ so that for all $\phi_- \in \Sigma_{scat}$ with $||\phi_-|| \leq \eta_o$, the equation

$$\phi(t) = e^{-itA}\phi_- + \int_{-\infty}^{t} e^{-iA(t-s)}J(\phi(s))ds \tag{60}$$

has a unique global continuous $\mathcal{H}$-valued solution $\phi(\cdot)$ with $|||\phi(\cdot)||| \leq 2\eta_o$. Moreover,

(a) For each t, $\phi(t) \in \Sigma_{scat}$

(b) $||\phi(t) - e^{-itA}\phi_-|| \to o$ as $t \to -\infty$

(c) In case $q > 1$, (b) can be strengthened to:

$$||e^{-itA}\phi(t) - \phi_-||_{scat} \to o \quad \text{as } t \to -\infty$$

Proof. The basic idea is to use the contraction mapping method which we employed in Section 1 except with initial conditions at $t = -\infty$. Let $X(\eta,\phi_-)$ denote the set of continuous $\mathcal{H}$-valued functions $\psi(t)$ so that $|||\psi(t) - e^{-itA}\phi_-||| \leq \eta$. Assume that $||\phi_-||_{scat} \leq \eta \leq \delta$ where δ is chosen so that hypothesis (iii) holds, and for $\psi(\cdot) \in X(\eta,\phi_-)$ define

$$(\mathcal{J}\psi)(t) = \int_{-\infty}^{t} e^{-iA(t-s)}J(\psi(s))ds \tag{61}$$

As in the proof of Theorem 1, it is easy to check that $e^{-iA(t-s)}J(\psi(s))$ is a continuous function of s for each t. Further, since

$$|||\psi(t)||| \leq |||e^{-itA}\phi_-||| + \eta \leq 2\eta$$

we have,

$$||J(\psi(s))|| \le \beta\ ||\psi(s)||_a^q\ ||\psi(s)|| \qquad \text{by (57)}$$

$$\le \beta(2\eta)^{q+1}(1+|s|)^{-dq}$$

so, the right hand side of (61) makes sense and

$$||(\mathcal{J}\psi)(t)|| \le \int_{-\infty}^{t} ||e^{-iA(t-s)}J(\psi(s))||ds$$

$$\le \beta(2\eta)^{q+1}\int_{-\infty}^{t}(1+|s|)^{-dq}ds \qquad (62)$$

$$< \infty$$

since $dq > 1$. Also,

$$||(\mathcal{J}\psi)(t)||_a \le \int_{-\infty}^{t} ||e^{-iA(t-s)}J(\psi(s))||_a ds$$

$$\le c_2\int_{-\infty}^{t}(1+|t-s)^{-d}(||J(\psi(s))||_b + ||J(\psi(s)||)ds \qquad \text{by (59)}$$

$$\le c_2\beta\int_{-\infty}^{t}(1+|t-s|)^{-d}\{\ ||\psi(s)||_a^q(1+\ 2||\psi(s)||)\}\ ds \qquad \text{by (57)}$$

$$\le c_2\beta(2\eta)^q(1+2\eta)\int_{-\infty}^{t}(1+|t-s|)^{-d}(1+|s|)^{-dq}\ ds \qquad (63)$$

$$\le c_2\beta(2\eta)^q(1+2\eta)c_3(1+|t|)^{-d}$$

by the lemma proved after the completion of this proof. Therefore $|||(\mathcal{J}\psi)(t)||| < \infty$. We now define

$$(M\psi)(t) = e^{-iAt}\phi_- + (\mathcal{J}\psi)(t)$$

and choose η_o (and δ in the case $q = 1$) small enough so that

$$\beta(2\eta)^{q+1}\int_{-\infty}^{\infty}(1+|s|)^{-dq}ds \le \eta_o/2 \qquad (64)$$

$$c_2\beta(2\eta_o)^q(1+2\eta_o)c_3 \leq \eta_o/2$$

It is easy to check that $(M\psi)(t)$ is continuous. Thus, for $\eta \leq \eta_o$, M maps $X(\eta,\phi_-)$ into itself. Further, it is easy to check using (58) that by choosing η_o still smaller (or δ in the case $q = 1$) we can guarantee that M is a contraction. Thus, since $X(\eta,\phi_-)$ is a complete metric space, M has a unique fixed point $\phi(\cdot)$ in $X(\eta,\phi_-)$. By the definition of M, $\phi(\cdot)$ is a global solution of (60). Notice also that $|||\phi(\cdot)||| \leq 2\eta_o$.

To prove uniqueness, let $\phi_1(t)$ be another solution of (60) with $|||\phi_1(t)||| < \infty$. Then

$$||\phi(t)-\phi_1(t)|| \leq \int_{-\infty}^{t} ||J(\phi(s)) - J(\phi_1(s))||ds$$

$$\leq \int_{-\infty}^{t} \beta(||\phi(s)||_a + ||\phi_1(s)||_a)^q \, ||\phi(s) - \phi_1(s)||ds$$

$$\leq \beta(|||\phi(\cdot)||| + |||\phi_1(\cdot)|||)^q \big(\sup_{-\infty<s\leq t} ||\phi(s)-\phi_1(s)||\big) \int_{-\infty}^{t} (1+|s|)^{-dq}ds$$

so

$$\sup_{-\infty<s\leq t} ||\phi(s) - \phi_1(s)||$$

$$\leq \{\beta(|||\phi(\cdot)||| + |||\phi_1(\cdot)|||)^q \int_{-\infty}^{t} (1+|s|)^{-dq}ds\} \sup_{-\infty<s\leq t} ||\phi(s) - \phi_1(s)||$$

But this gives a contradiction for t sufficiently close to $-\infty$ unless $\phi(s) = \phi_1(s)$ for all $s \leq t$. By local uniqueness (see Theorem 1) $\phi(s) = \phi_1(s)$ for all s.

To show that $\phi(t) \in \sum_{\text{scat}}$ for each t; we fix t and compute:

$$\sup_r ||e^{-irA}\phi(t)|| \leq \sup_r ||e^{-irA}e^{-itA}\phi|| + \int_{-\infty}^{t} ||e^{-iA(t+r-s)}J(\phi(s))||ds$$

$$\leq \sup_r ||e^{-irA}\phi_-|| + \beta \int_{-\infty}^{t} ||\phi(s)||_a^q \, ||\phi(s)||ds$$

$$\leq ||\phi_-||_{scat} + \frac{\eta_o}{2}$$

$$\sup_r (1+|r|)^d ||e^{-irA}\phi(t)||_a \leq \sup_r (1+|r|)^d ||e^{-i(t+r)A}\phi_-||_a$$

$$+ \sup_r (1+|r|)^d \int_{-\infty}^{t} ||e^{-iA(t+r-s)}J(\phi(s))||_a ds$$

$$\leq \sup_r \{(1+|r|)^d (1+|t+r|)^{-d} ||\phi_-||_{scat}\}$$

$$+ \sup_r \{(1+|r|)^d \int_{-\infty}^{t} (1+|t+r-s|)^{-d} (1+|s|)^{-dq} ds\} \; \beta c_2 (1+2\eta_o)(4\eta_o)^q$$

$$\leq \sup_r \{(1+|r|)^d (1+|t+r|)^{-d}\} (||\phi_-||_{scat} + \frac{\eta_o}{2})$$

Thus, $\phi(t) \in \sum_{scat}$.

To prove (b) and (c), we estimate:

$$||e^{-irA}(e^{itA}\phi(t)-\phi_-)|| \leq \int_{-\infty}^{t} ||e^{-irA}e^{isA}J(\phi(s))||ds$$

$$\leq \beta \int_{-\infty}^{t} ||\phi(s)||_a^q \, ||\phi(s)||ds$$

$$\leq \beta (2\eta_o)^q \eta_o \int_{-\infty}^{t} (1+|s|)^{-dq} ds$$

$$\longrightarrow o \quad \text{as} \quad t \to -\infty$$

Taking $r = o$, this proves (b). Further,

$$(1+|r|)^d ||e^{-irA}(e^{itA}\phi(t)-\phi_-)||_a$$

$$\leq (1+|r|)^d \int_{-\infty}^{t} ||e^{-iA(r-s)}J(\phi(s))||_a ds$$

$$\leq (1+|r|)^d \int_{-\infty}^{t} c_2(1+|r-s|)^{-d} \{||J(\phi(s))|| + ||J(\phi(s))||_b\} ds$$

$$\leq (1+r)^d c_2 \int_{-\infty}^{t} (1+|r-s|)^{-d}\beta\{2||\phi(s)||_a^q||\phi(s)|| + ||\phi(s)||_a^q\} ds$$

$$\leq \beta\ c_2(2\eta_0^{q+1}+ \eta_0^q) \int_{-\infty}^{t} (1+|r|)^d(1+|r-s|)^{-d}(1+|s|)^{-dq} ds$$

So, if $q > 1$,

$$\sup_r (1+|r|)^d ||e^{-irA}(e^{itA}\phi(t) - \phi_-)||_a \longrightarrow 0$$

as $t \to -\infty$ by part (b) of the lemma below. From this and the above it follows that

$$||e^{itA}\phi(t) - \phi_-||_{scat} \to 0$$

as $t \to -\infty$. ∎

Notice that the solution of (60) constructed above satisfies

$$\phi(t) = e^{-iAt}\phi_- + \int_{-\infty}^{t} e^{-iA(t-s)}J(\phi(s))ds$$

$$= e^{-iAt}\ \{\phi_- + \int_{-\infty}^{0} e^{-iAs}J(\phi(s))ds\} + \int_{0}^{t} e^{-iA(t-s)}J(\phi(s))ds$$

so $\phi(t)$ satisfies (52) with

$$\phi_0 = \phi_- + \int_{-\infty}^{0} e^{-iAs}J(\phi(s))ds$$

The following lemma completes the proof of Theorem 16.

Lemma (a) Suppose that $q \geq 1$, $d > 0$, and $dq > 1$. Then

$$\int_{-\infty}^{\infty} (1+|t-s|)^{-d}(1+|s|)^{-dq}\, ds \leq c(1+|t|)^{-d}$$

(b) Suppose that $q > 1$, $d > 0$, and $dq > 1$. Then

$$\sup_{r} \{(1+|r|)^{d} \int_{t_1}^{t_2} (1+|r-s|)^{-d}(1+|s|)^{-dq} ds\} \longrightarrow 0$$

as $t_1, t_2 \longrightarrow +\infty$ or $t_1, t_2 \longrightarrow -\infty$.

Proof. To prove (a) it is sufficient to consider the case where t is large and positive. The proof for t large and negative is similar. We break the integral into two parts and estimate:

$$\int_{|s-t|\geq\frac{t}{2}} (1+|t-s|)^{-d}(1+|s|)^{-dq} ds \leq (1+\tfrac{t}{2})^{-d} \int_{|s-t|\geq\frac{t}{2}} (1+|s|)^{-dq} ds$$

$$\leq c(1+t)^{-d}\int_{-\infty}^{\infty} (1+|s|)^{-dq} ds$$

and for $d \neq 1$,

$$\int_{\frac{t}{2}}^{3\frac{t}{2}} (1+|t-s|)^{-d}(1+|s|)^{-dq} ds \tag{65}$$

$$\leq (1+\tfrac{t}{2})^{-dq} \{\int_{\frac{t}{2}}^{t} (1+(t-s))^{-d} ds + \int_{t}^{3\frac{t}{2}} (1+s-t)^{-d} ds\}$$

$$\leq 2(1+\tfrac{t}{2})^{-dq} \{(1-d)^{-1}((1+\tfrac{t}{2})^{-d+1}+1)\}$$

$$\leq c(1+t)^{-dq-d+1} + c(1+t)^{-dq}$$

$$\leq c(1+t)^{-d}$$

since $dq > 1$ and $q \geq 1$.

If $d = 1$, the right hand side of (65) can be estimated by

$$\leq 2(1+\tfrac{t}{2})^{-q} \log(1+\tfrac{t}{2})$$

$$\leq c(1+t)^{-1}$$

since $q > 1$ if $d = 1$. Combining these estimates proves (a).

We will prove (b) in the case $t_1, t_2 \to +\infty$. We assume that t_1 is large and positive and divide the domain of integration into two parts.

$$(1+|r|)^d \int_{[t_1,t_2]\setminus[\frac{r}{2},\frac{3r}{2}]} (1+|r-s|)^{-d}(1+|s|)^{-dq} ds$$

$$\leq (1+|r|)^d (1+|\tfrac{r}{2}|)^{-d} \int_{[t_1,t_2]\setminus[\frac{r}{2},\frac{3r}{2}]} (1+|s|)^{-dq} ds$$

$$\leq c\int_{t_1}^{t_2} (1+|s|)^{-dq} ds \tag{66}$$

and assuming $d \neq 1$,

$$(1+|r|)^d \int_{[t_1,t_2]\cap[\frac{r}{2},\frac{3r}{2}]} (1+|r-s|)^{-d}(1+|s|)^{-dq} ds$$

$$\leq (1+|r|)^d (1+|\tfrac{r}{2}|)^{-dq} \int_{[t_1,t_2]\cap[\frac{r}{2},\frac{3r}{2}]} (1+|r-s|)^{-d} ds$$

$$\leq (1+|r|)^d (1+|\tfrac{r}{2}|)^{-dq} \left[c+c(1+|r|)^{-d+1}\right]$$

$$\leq c_1 (1+|r|)^{d(1-q)} + c_2 (1+|r|)^{1-dq}$$

$$\leq c_3 (1+|t_1|)^{d(1-q)} + c_4 (1+|t_1|)^{1-dq}$$

since the integral is zero unless $r \geq \frac{2}{3} t_1$.

Combining this estimate and (66) we see that (b) holds since $dq > 1$, $q > 1$. The case $d = 1$ is handled as in the proof of (a). ∎

The fact that part (b) of the lemma holds only if $q > 1$ makes the case $q = 1$ special in all that follows. To see that the conclusion is actually false in the case $q = 1$, notice that

$$\int_{t_1}^{t_1+1} (1+|t-s|)^{-d} (1+|s|)^{-d} ds$$

$$\geq (1+|t_1+1|)^{-d} \int_{t_1}^{t_1+1} (1+|t-s|)^{-d} ds$$

$$= (1+|t_1+1|)^{-d} \int_0^1 (1+|s|)^{-d} ds$$

if $t = t_1$. Thus,

$$\sup_t (1+|t|)^d \int_{t_1}^{t_1+1} (1+|t-s|)^{-d} (1+|s|)^{-d} ds$$

$$\geq \left(\frac{1+|t_1|}{1+|t_1+1|}\right)^d \int_0^1 (1+|s|)^{-d} ds$$

We now have global existence and the right properties at $-\infty$. To construct the scattering operator we must construct a ϕ_+ so that $\phi(t) - e^{-iAt}\phi_+ \longrightarrow 0$ as $t \longrightarrow +\infty$.

Theorem 17 (the scattering operator for small data)
Assume all the hypotheses of Theorem 16 and let $\phi(t)$ be the solution of (60) corresponding to $\phi_- \in \Sigma_{scat}$ with $||\phi_-||_{scat} \leq \eta_o$. Then, for η_o sufficiently small,

(a) There exists $\phi_+ \in \Sigma_{scat}$, with $||\phi_+||_{scat} \leq 2\eta_o$, so that

$$||\phi(t) - e^{-itA}\phi_+|| \longrightarrow o \quad \text{as} \quad t \to +\infty$$

(b) The map $\phi_- \xrightarrow{S} \phi_+$ is a one to one and continuous (in the $||\ ||$ topology) map of the ball $\{\psi \in \Sigma_{scat} \,|\, ||\psi||_{scat} \leq \eta_o\}$ into the ball $\{\psi \in \Sigma_{scat} \,|\, ||\psi||_{scat} \leq 2\eta_o\}$.

Except in the case $q = 1$, the following also hold:

(c) $$||e^{itA}\phi(t) - \phi_+||_{scat} \longrightarrow o \quad \text{as } t \to +\infty$$

(d) S is continuous in the $||\cdot||_{scat}$-topology.

Proof. From Theorem 16 we know that $|||\phi(t)||| \leq 2\eta_o$. Thus,

$$||e^{it_1A}\phi(t_1) - e^{it_2A}\phi(t_2)|| \leq || \int_{t_1}^{t_2} e^{isA}J(\phi(s))ds||$$

$$\leq \int_{t_1}^{t_2} \beta||\phi(s)||_a^q \, ||\phi(s)||ds$$

$$\leq \beta(2\eta_o)^{q+1} \int_{t_1}^{t_2} (1+|s|)^{-dq}ds$$

by (57). Thus $\{e^{itA}\phi(t)\}$ is Cauchy in $\mathcal{H}$ as $t \to +\infty$ since $dq > 1$. Letting

$$\phi_+ = \lim_{t\to+\infty} e^{itA}\phi(t)$$

we have

$$||\phi(t) - e^{-itA}\phi_+|| \to o \quad \text{as } t \to +\infty$$

by the unitarity of e^{-itA}. To show that $\phi_+ \in \Sigma_{scat}$, observe that

$$e^{itA}\phi(t) = \phi_- + \int_{-\infty}^{t} e^{isA}J(\phi(S))ds$$

Letting $t \to +\infty$ we conclude that

$$\phi_+ = \phi_- + \int_{-\infty}^{\infty} e^{isA}J(\phi(s))ds$$

Now, by (59) and (57),

$$||e^{-iA(t-s)}J(\phi(s))||_a \leq c_2(1+|t-s|)^{-d}\{||J(\phi(s))|| + ||J(\phi(s))||_b\}$$

$$\leq c_2 \ \beta(1+|t-s|)^{-d}\{||\phi(s)||_a^q \ (1+2||\phi(s)||)\}$$

$$\leq c_2 \ \beta(2\eta_0)^q(1+\eta_0)(1+|t-s|)^{-d}(1+|s|)^{-dq}$$

for each s and t. Since for fixed t,

$$e^{-itA}\phi_+ = e^{-itA}\phi_- + \int_{-\infty}^{\infty} e^{-iA(t-s)}J(\phi(s))ds$$

we conclude that $||e^{-itA}\phi_+||_a < \infty$ and

$$\sup_t(1+|t|)^d||e^{-itA}\phi_+||_a$$

$$\leq \sup_t(1+|t|)^d||e^{-itA}\phi_-||_a$$

$$+ c_2\beta(2\eta_0)^q(1+4\eta_0)\sup_t\{(1+|t|)^d \int_{-\infty}^{\infty}(1+|t-s|)^{-d}(1+|s|)^{-dq}ds\}$$

$$\leq \sup_t (1+|t|)^d ||e^{-itA}\phi_-||_a + \frac{\eta_o}{2}$$

by the lemma (part a) and the choice of η_o in Theorem 16. Thus

$$||\phi_+||_{scat} \leq ||\phi_-||_{scat} + \frac{\eta_o}{2} + \frac{\eta_o}{2} \leq 2\eta_o$$

This proves (a).

We can now define S: $\phi_- \to \phi_+$ and it is clear that S takes $\{\phi \in \Sigma_{scat} \mid ||\phi||_{scat} \leq \eta_o\}$ into $\{\psi \in \Sigma_{scat} \; ||\psi||_{scat} \leq 2\eta_o\}$. The proof that S is one to one is similar to the uniqueness proof in Theorem 16. To prove that S is continuous in the $||\cdot||$-topology we proceed as follows. Let $\phi_-^{(1)}$ and $\phi_-^{(2)}$ be in $\{\psi| \; ||\psi||_{scat} \leq \eta_o\}$ and let $\phi^{(1)}(t)$ and $\phi^{(2)}(t)$ be the corresponding solutions of *(60)*. Then

$$||\phi^{(1)}(t)-\phi^{(2)}(t)|| \leq ||\phi_-^{(1)} - \psi_-^{(2)}|| + \int_{-\infty}^{t} ||J(\phi^{(1)}(s)) - J(\phi^{(2)}(s))||ds$$

$$\leq ||\phi_-^{(1)} - \phi_-^{(2)}|| + \beta(|||\phi^{(1)}(s)||| + |||\phi^{(2)}(s)|||)^q \int_{-\infty}^{t} (1+|s|)^{-dq} ||\phi^{(1)}(s)-\phi^{(2)}(s)||ds$$

$$\leq ||\phi_-^{(1)}-\phi_-^{(2)}|| + \beta(4\eta_o)^q \int_{-\infty}^{t} (1+|s|)^{-dq} ||\phi^{(1)}(s)-\phi^{(2)}(s)||ds$$

Since $dq > 1$, interation of this inequality implies that

$$||\phi^{(1)}(t)-\phi^{(2)}(t)|| \leq ||\phi_-^{(1)} - \phi_-^{(2)}|| \exp\{\int_{-\infty}^{t} \beta(4\eta_o)^q (1+|s|)^{-dq} ds\}$$

By part (a),

$$e^{iAt}(\phi^{(1)}(t) - \phi^{(2)}(t)) \xrightarrow{||\cdot||} \phi_+^{(1)} - \phi_+^{(2)}$$

as $t \to \infty$, so we conclude that

$$||\phi_+^{(1)} - \phi_+^{(2)}|| \leq c(\eta_0)||\phi_-^{(1)} - \phi_-^{(2)}|| \tag{67}$$

which proves that S is continuous in the $||\cdot||$ norm.

To prove (c) we estimate:

$$(1+|r|)^d||e^{-irA}(e^{it_2A}\phi(t_2) - e^{it_1A}\phi(t_1)||_a$$

$$\leq (1+|r|)^d \int_{t_1}^{t_2} ||e^{-iA(r-s)}J(\phi(s))||_a ds$$

$$\leq (1+|r|)^d c_2\beta\int_{t_1}^{t_2}(1+|r-s|)^{-d}(||\phi(s)||_a^q + 2||\phi(s)||_a^q\ ||\phi(s)||)ds$$

$$\leq c_2\beta(2\eta_0)^q(1+2\eta_0)\ \{(1+|r|)^d \int_{t_1}^{t_2}(1+|r-s|)^{-d}(1+|s|)^{-dq}ds\}$$

By part (b) of the lemma, the sup of the right hand side goes to zero as $t_1, t_2 \to +\infty$ if $q > 1$. It follows in this case that $e^{itA}\phi(t)$ is Cauchy in the $||\cdot||_{scat}$ norm so $||e^{itA}\phi(t) - \phi_+||_{scat} \to 0$ as $t \to +\infty$

To prove (d) we use (c) and the continuity already proven in (67) Let $\phi^{(1)}(t), \phi^{(2)}(t)$ be as in the proof of (b) and define:

$$Q(s) = (1+|s|)^d||\phi^{(1)}(s) - \phi^{(2)}(s)||_a$$

and

$$P(t) = \sup_{-\infty\leq s\leq t} Q(s)$$

Then,

$$Q(t) \leq (1+|t|)^d||e^{-itA}(\phi_-^{(1)} - \phi_-^{(2)})||$$

$$+ (1+|t|)^d \int_{-\infty}^{t}||e^{-iA(t-s)}(J(\phi^{(1)}(s)) - J(\phi^{(2)}(s))||_a ds$$

$$\leq ||\phi_-^{(1)} - \phi_-^{(2)}||_{scat} + \beta c_2 (1+|t|)^d$$

$$\cdot \int_{-\infty}^{t} (1+|t-s|)^{-d} (||\phi^{(1)}(s)||_a + ||\phi^{(2)}(s)||_a)^{q-1} ||\phi^{(1)}(s) - \phi^{(2)}(s)||_a ds$$

$$+ 2\beta c_2 (1+|t|)^d \int_{-\infty}^{t} (1+|t-s|)^{-d} (||\phi^{(1)}(s)||_a + ||\phi^{(2)}(s)||_a)^q ||\phi^{(1)}(s) - \phi^{(2)}(s)|| ds$$

$$\leq ||\phi_-^{(1)} - \phi_-^{(2)}||_{scat} + \beta c_2 (4\eta_0)^{q-1} (1+|t|)^d \int_{-\infty}^{t} (1+|t-s|)^{-d} (1+|s|)^{-dq} Q(s) ds$$

$$+ 2\beta c_2 (4\eta_0)^q c(\eta_0) ||\phi_-^{(1)} - \phi_-^{(2)}||_{scat} (1+|t|)^d \int_{-\infty}^{t} (1+|t-s|)^{-d} (1+|s|)^{-dq} ds$$

where we have used (67) in the last step. Thus, using part (a) of the lemma we see that

$$Q(t) \leq c_4 ||\phi_-^{(1)} - \phi_-^{(2)}||_{scat} + \beta c_5 (\sup_{-\infty \leq s \leq t} Q(s))$$

or

$$P(t) \leq c_4 ||\phi_-^{(1)} - \phi_-^{(2)}||_{scat} + \beta c_5 P(t) \qquad (68)$$

where the constants c_4 and c_5 just depend on η_0 and βc_5 may be made arbitrarily small by choosing η_0 small in the case $q > 1$ and β small in the case $q = 1$. Thus, for η_0 small enough

$$P(t) \leq \frac{c_4}{(1-\beta c_5)} ||\phi_-^{(1)} - \phi_-^{(2)}||_{scat}$$

Combining this with (67) we conclude that there is a constant c_6 so that:

$$|||\phi^{(1)}(t) - \phi^{(2)}(t)||| \leq c_6 ||\phi_-^{(1)} - \phi_-^{(2)}||_{scat} \qquad (69)$$

Finally, we estimate

$$(1+|r|)^d ||e^{-irA}(e^{iAt}\phi^{(1)}(t) - e^{iAt}\phi^{(2)}(t))||_a$$

$$\leq ||\phi_-^{(1)} - \phi_-^{(2)}||_{scat} + (1+|r|)^d \int_{-\infty}^t ||e^{-iA(r-s)}(J(\phi^{(1)}(s)) - J(\phi^{(2)}(s)))||_a ds$$

$$\leq c_4 ||\phi_-^{(1)} \phi_-^{(2)}||_{scat} + c_7 |||\phi^{(1)}(t) - \phi^{(2)}(t)|||(1+|r|)^d \int_{-\infty}^t (1+|r-s|)^{-d}(1+|s|)^{-dq} ds$$

$$\leq c_8 ||\phi_-^{(1)} - \phi_-^{(2)}||_{scat}$$

by the usual estimates and (69). From this and (67) we have

$$||e^{iAt}\phi^{(1)}(t) - e^{iAt}\phi^{(2)}(t)||_{scat} \leq c||\phi_-^{(1)} - \phi_-^{(2)}||_{scat} \tag{70}$$

where c depends on η_0 (and δ if $q = 1$) but not on t or $\phi_-^{(1)}$, $\phi_-^{(2)}$. From part (c),

$$e^{iAt}\phi^{(1)}(t) \xrightarrow{||\cdot||_{scat}} \phi_+^{(1)}$$
$$e^{iAt}\phi^{(2)}(t) \xrightarrow{||\cdot||_{scat}} \phi_+^{(2)} \tag{71}$$

if $q > 1$, so in that case we conclude that

$$||\phi_+^{(1)} - \phi_+^{(2)}||_{scat} \leq c||\phi_-^{(1)} - \phi_-^{(2)}||_{scat}$$

which proves (d). ∎

Notice that we have the following interesting situation. For all $q \geq 1$, $\phi_+^{(1)}$ and $\phi_+^{(2)}$ are in Σ_{scat} by part (a) of the theorem and for all $q \geq 1$ we have uniform continuity of the map $\phi_- \to e^{iAt}\phi(t)$ in the $||\cdot||_{scat}$ norm for all finite times t (this is just (70)). But it is only in the case $q > 1$ that we can conclude from this that S is continuous in the $||\cdot||_{scat}$ norm because only in this case does (71) hold.

We give applications of these theorems in Section 11. But before going on let me emphasize again two very important aspects of Theorems 16 and 17. Namely, the hypotheses of the theorems did not require _a priori_ estimates on solutions of the non-linear equation nor were any energy inequalities used. The only requirement was that the solutions of the _linear_ equation decay sufficiently rapidly and that the non-linearity be of sufficiently high degree. In particular, the method will work for cases where the conserved energy is not bounded below.

The idea that one can develop a scattering theory for "small data" goes back to the paper by Segal [33]. In this paper Segal concentrates on applications to the Klein-Gordon equation with u^p interaction and many special properties of the kernal of $e^{-iA(t-s)}$ are exploited. Strauss [38] simplified Segal's work and formulated the problem in terms of abstract hypotheses on A and J. We have followed the elaboration of Strauss' ideas in [27].

9. Global existence for small data

There is another aspect of Theorems 16 and 17 which is so important that it deserves to be set out separately. That is, if we have the hypotheses (i), (ii), (iii), then the initial value problem at $t = o$, namely,

$$(*) \qquad \begin{aligned} \phi(t) &= e^{-iAt}\phi_o + \lambda\int_o^t e^{-iA(t-s)}J(\phi(s))ds \\ \phi(o) &= \phi_o \end{aligned}$$

has a global solution if $||\phi_o||_{scat}$ is small enough. The proofs of this and the other parts of the theorem below are almost exactly the same as the proofs of Theorem 16 and 17.

Theorem 18 (global existence for small data) Let A be a self-adjoint operator on a Hilbert space $\mathcal{H}$ and J a non-linear mapping of $\mathcal{H}$ into itself. Suppose that there exist norms $||\cdot||_a$, $||\cdot||_b$ so that the hypotheses (i), (ii), (iii) of Section 8 hold. Let Σ_{scat} be as defined in Section 8. Then,

(a) For each $\phi_o \in \Sigma_{scat}$, the equation (*) has a global continuous Σ_{scat}-valued solution $\phi(t)$ if either $||\phi_o||_{scat}$ or λ is small enough.

(b) Let λ be fixed. Then there exists $\eta_o > o$ so that the global existence of part (a) holds for all ϕ_o satisfying $||\phi_o||_{scat} \leq \eta_o$. Further, for each such ϕ_o, there exist ϕ_- and ϕ_+ in Σ_{scat} so that

$$||\phi(t) - e^{-iAt}\phi_-|| \longrightarrow o \qquad \text{as} \quad t \longrightarrow -\infty$$

$$||\phi(t) - e^{-iAt}\phi_+|| \longrightarrow o \qquad \text{as} \quad t \longrightarrow +\infty$$

and the maps

$$\phi(o) \xrightarrow{(\Omega_-)^{-1}} \phi_- \quad , \quad \phi(o) \xrightarrow{(\Omega_+)^{-1}} \phi_+$$

are one to one and continuous in the $||\cdot||$ norm.

(c) If $q > 1$ (in hypothesis (iii)), then (b) can be strengthened to

$$||e^{iAt}\phi(t) - \phi_-||_{scat} \longrightarrow o \quad \text{as} \quad t \longrightarrow -\infty$$

$$||e^{iAt}\phi(t) - \phi_+||_{scat} \longrightarrow o \quad \text{as} \quad t \longrightarrow +\infty$$

and $(\Omega_+)^{-1}$, $(\Omega_-)^{-1}$ are continuous in the $|| \ ||_{scat}$ norm.

This theorem can be used to show that for small enough initial data

$$u_{tt} - \Delta u + m^2 u = \lambda u^p \qquad x \in R^3$$

has global strong solutions as long as p is large enough no matter what the sign of λ is (see Section 11, part c).

10. Existence of the Wave operators

In the case where global solutions of (52) exist for all initial data, we can use the ideas of the last section to construct the wave operators on all of Σ_{scat}, not just on a small ball. As before we will denote by M_t the map

$$M_t : \phi(o) \longrightarrow \phi(t)$$

where $\phi(t)$ is the local solution of (52). The idea is as follows. The crucial estimates in the existence proof of Theorem 16 were (62) and (63). In order to make

$$(M\psi)(t) = e^{-iAt}\phi_- + (\mathcal{J}\psi)(t)$$

a contraction mapping we had to make the right sides of these estimates small and we did this by choosing η small enough. If we are willing to restrict $\psi(t)$ to an interval $(-\infty, T_o)$ where T_o is close to $-\infty$ then (if $q > 1$) the right sides will be small because the integrals are small even if η is not small. In this way we get a solution on $(-\infty, T_o)$ with the right properties. Then local uniqueness and global existence allow one to extend this solution to $(-\infty, 0]$ and thus define the wave operator Ω_-. Since the details are similar to those in Theorems 16 and 17, we will just sketch the proof. For convenience we restate the basic integral equation

$$\phi(t) = e^{-itA}\phi(s) + \int_o^t e^{-iA(t-s)} J(\phi(s))ds \tag{72}$$

Theorem 19 (existence of the wave operators)
Let A be a self-adjoint operator on a Hilbert space $\mathcal{H}$ and J a non-linear mapping of $\mathcal{H}$ into itself. Suppose that there exist norms $||\cdot||_a, ||\cdot||_b$ so that the hypotheses (i), (ii) and (iii) hold with $q > 1$ (see Section 8). Suppose that for each η, T the solutions $M_t\phi(o)$ of (72) are uniformly bounded (in $||\cdot||$-norm) for all $||\phi(o)|| \leq \eta$ and all $o < |t| \leq T$. Then,

(a) For each $\phi_- \in \Sigma_{scat}$ there is a unique global solution $\phi(t)$ of (72) so that $\phi(t) \in \Sigma_{scat}$ for each t, and

$$||\phi(t) - e^{-iAt}\phi_-|| \to o \quad \text{as } t \to -\infty$$

$$||e^{iAt}\phi(t) - \phi_-||_{scat} \to o \quad \text{as } t \to -\infty$$

(b) The mapping $\Omega_- : \phi_- \to \phi(o)$ is a one to one map of Σ_{scat} into Σ_{scat} which is uniformly continuous on balls in Σ_{scat}.

(c) Analogous statements as in (a) and (b) hold for $\phi_+ \in \Sigma_{scat}$, $t \to +\infty$, and the map $\Omega_+ : \phi_+ \to \phi(o)$.

Proof. Let $\phi_- \in \Sigma_{scat}$ be given, with $||\phi_-||_{scat} \leq \eta$ (note that η is not assumed small). Let $X(\eta,\phi_-,T)$ denote the $\mathcal{H}$-valued continuous functions $\phi(\cdot)$ on $(-\infty,T]$ so that

$$|||\psi(t) - e^{-itA}\phi_-|||_{(-\infty,T]}$$

$$\equiv \sup_{-\infty \leq t \leq T} ||\psi(t)-e^{-itA}\phi_-|| + \sup_{-\infty \leq t \leq T} (1+|t|)^d ||\psi(t) - e^{-itA}\phi_-||_a$$

$$\leq \eta$$

For each T, $X(\eta,\phi_-,T)$ is a complete metric space. We define $\mathcal{J}$ by (61) and M by

$$(M\psi)(t) = e^{-iAt}\phi_- + (\mathcal{J}\psi)(t)$$

The estimates (62) and (63) of Theorem 16 and the following lemma show that $|||\mathcal{J}(\psi)(t)|||_{(-\infty,T]}$ can be made small by choosing T sufficiently close to $-\infty$, so in that case M will take $X(\eta,\phi_-,T)$ into itself. The same sort of estimates show that M is a contraction for T close to $-\infty$ and so M has a fixed point $\phi(\cdot)$ in $X(\eta,\phi_-,T_o)$ for some fixed T_o. Notice that we need $q > 1$ since we need part (b) of the lemma to prove the smallness. Just as in Theorem 16 one can prove that $\phi(t) \in \Sigma_{scat}$ for each $t \in (-\infty,T_o]$ and that the limits in (a) hold.

By the estimates (62) and (63) we can use the same T_o for all

ϕ_- with $||\phi_-||_o \le \eta$. Thus, we can define the map

$$\Omega_+^{T_o} : \phi_- \to \phi(T_o)$$

on $\mathcal{B}_\eta \equiv \{\psi \in \Sigma_{scat} \,|\, ||\psi||_{scat} \le \eta\}$. By the same proofs as in Theorem 16 and 17, $\Omega_+^{T_o}$ is a one to one uniformly continuous map of $\mathcal{B}_\eta$ into Σ_{scat}.

It can easily be checked that for $t \le T_o$ our solution $\phi(t)$ satisfies

$$\phi(t) = e^{-iA(t-T_o)}\phi(T_o) + \int_{T_o}^{t} e^{-iA(t-s)} J(\phi(s))ds$$

Since J is Lipschitz by hypothesis (iii) this equation can be solved locally in a neighborhood of T_o by $\phi(t) = M_{t-T}\,\phi(T_o)$ and by the boundedness hypothesis on M_t this solution is global in t. By local uniqueness this definition of $\phi(t)$ coincides with the $\phi(t)$ defined earlier for $t \le T_o$. It is easy to check that $\phi(t)$ satisfies (72) and that (57),(58), and (59) imply that $\phi(t)$ is in Σ_{scat} for all t. In fact, the uniform boundedness of M_t assumed in the hypothesis implies by Theorem 14 that, for each t, M_t is uniformly continuous on balls in $\mathcal{H}$. This combined with (58) and (59) can be used to easily prove that M_t is a uniformly continuous one to one map on balls in Σ_{scat} for each fixed t. Now, for each η we define

$$\Omega_+ = M_{-T_o(\eta)}\Omega_+^{T_o(\eta)}$$

That is, $$\Omega_+ : \phi_- \longrightarrow \phi(o)$$

By the properties of M_{-T} and $\Omega_+^{T_o(\eta)}$, we have that Ω_+ is a one to one uniformly continuous map of $\mathcal{B}_\eta$ into Σ_{scat}. Since η was arbitrary, Ω_+ takes Σ_{scat} into Σ_{scat} and is uniformly continuous on balls. An easy argument shows that Ω_+ is one to one on all of Σ_{scat}. This proves (a) and (b); the proof of (c) is similar. ∎

This theorem shows that when we have the hypotheses (i), (ii), (iii) on J and global existence then we can construct the wave operators on Σ_{scat}. The existence of the scattering operator depends on whether

$$\text{range } \Omega_+ = \text{range } \Omega_-$$

so that we can define $S = \Omega_+^{-1}\Omega_-$. This (much more difficult) problem is discussed in Section 12. We remark that in applications not all the hypotheses (i), (ii), and (iii) may be satisfied. Often one can construct the wave operators anyway by using special properties of the particular equations and the <u>idea</u> of Theorem 19. We will see an example of this in Section 11c.

11. Applications

In this lecture we will sketch some applications of the theorems in sections 8,9,10. In most cases we will not give very many details. Our main purpose is to show how one chooses the norms $||\cdot||$, $||\cdot||_a$, and $||\cdot||_b$ in a variety of situations. The typical procedure is as follows. The norm $||\cdot||_a$ is a sup norm on some or all of the components. The norm $||\cdot||_b$ is then the best (i.e. lowest) norm for which one can prove the decay estimate (ii). Then $||\cdot||$ is chosen to be a Hilbert space norm which is as simple as possible so that (i) holds. This is usually done by a Sobolev inequality. Thus, the three norms $||\cdot||$, $||\cdot||_a$, $||\cdot||_b$ are typically determined from the linear problem alone. One then checks to see what properties the non-linear term J must have so that (iii) holds. Of course, if the J in which one is interested doesn't satisfy (iii) one may be able to go back and change the norms $|| \; ||, || \; ||_a, || \; ||_b$ so that it does.

In all our examples we treat $B = \sqrt{-\Delta + m^2}$ as though it acted on powers of functions like differentiation. The correct details can be easily supplied as in part A of section 2.

Part A. The non-linear Schrödinger equation

We begin with an easy example, the non-linear Schrödinger equation in one dimension,

$$u_t - iu_{xx} + ku^p = o \tag{73}$$

$$u(x,o) = f(x)$$

because it illustrates nicely the method for choosing the norms described above. The corresponding free equation is

$$u_t - iu_{xx} = o$$

$$u(x,o) = g(x)$$

with $A = -\frac{d^2}{dx^2}$. The solution can be written explicitly as

$$u(x,t) = (4\pi it)^{-1/2} \int e^{i(x-y)^2/4t} f(y)dy$$

and thus,

$$||u(x,t)||_\infty \leq \frac{t^{-1/2}}{4\pi} ||f||_1$$

Therefore, we choose

$$||u||_a = ||u||_\infty \quad , \quad ||u||_b = ||u||_1$$

We have thus satisfied hypothesis (ii) with $d = \frac{1}{2}$. Notice that we cannot choose $L^2(R)$ as our Hilbert space because it is not true that $||u||_\infty \leq C||u||_2$. However, in one dimension it is true that (see Theorem 4 in part c of section 2):

$$||u||_\infty \leq C||Bu||_2^{1/2} \, ||u||_2^{1/2} \leq C_1||Bu||_2 \tag{74}$$

where as usual $B = \sqrt{-\Delta + m^2}$.

Thus, we take

$$||u|| = ||Bu||_2$$

so that hypothesis (i) is satisfied. To check for which p (iii) holds, we compute,

$$\begin{aligned}
||J(u_1) - J(u_2)|| &= K||B(u_1^p - u_2^p)||_2 \\
&= Kp||(Bu_1)u_1^{p-1} - (Bu_2)u_2^{p-1}||_2 \\
&\leq Kp||(Bu_1)(u_1 - u_2)P(u_1,u_2)||_2 \\
&\quad + Kp||(Bu_1 - Bu_2)u_2^{p-1}||_2 \\
&\leq C||Bu_1||_2||u_1 - u_2||_\infty||P(u_1,u_2)||_\infty \\
&\quad + C||B(u_1 - u_2)||_2||u_2^{p-1}||_\infty
\end{aligned}$$

$$\leq C||Bu_1||_2 \; ||B(u_1 - u_2)||_2 \; (||u_1||_\infty + ||u_2||_\infty)^{p-2}$$

$$+ C||B(u_1 - u_2)||_2 \; ||u_2||_\infty^{p-1}$$

$$\leq \beta(||u_1||_a + ||u_2||_a)^{p-2} \; ||u_1 - u_2||$$

for $||u_1||$ and $||u_2||$ small. Thus, the first hypothesis in (iii) holds with $q = p-2$. Similarly,

$$||J(u_1) - J(U_2)||_b = ||u_1^p - u_2^p||_1$$

$$= ||(u_1 - u_2)Q(u_1,u_2)||_1$$

$$\leq C||B(u_1 - u_2)||_2(||u_1||_2+||u_2||)(||u_1||_\infty+||u_2||_\infty)^{p-2}$$

$$\leq \beta||u_1 - u_2||(||u_1||_\infty + ||u_2||_\infty)^{p-2}$$

so the second part of (iii) also holds with $q = p-2$. Now, since $d = \frac{1}{2}$ we must have $q > 2$ so that $dq > 1$. Therefore if $p > 4$, Theorems 16 and 17 provide small data global existence and a scattering theory for equation (73).

The application of these small data techniques to (73) for high p is due to Strauss [38].

Part B. $u_{tt} - u_{xx} + m^2u = \lambda u^p$, one dimension

In order to discuss the equation

$$u_{tt} - u_{xx} + m^2u = \lambda u^p \tag{75}$$

we first need a decay estimate for the linear equation:

$$u_{tt} - u_{xx} + m^2u = 0$$

$$u(x,0) = f \tag{76}$$

$$u_t(x,0) = g$$

Although this is a fact about a linear equation, the proof is non-trivial so we will provide a sketch.

Lemma 1 Suppose $f, g \in \mathcal{S}(R)$ and let $u(x,t)$ be the solution of (76) with initial data $\langle f, g\rangle$. Then

$$||u(x,t)||_\infty \leq Ct^{-1/2}\{||f||_1 + ||f'||_1 + ||f''||_1 + ||g'||_1 + ||g||_1\} \quad (77)$$

Proof For each t, $u(x,t) = u(t)$ is given by

$$u(t) = \cos(Bt)f + \frac{\sin(Bt)}{B} g$$

or

$$\widehat{u(t)} = \cos(\sqrt{k^2+m^2}t)\ \hat{f}(k) + \frac{\sin(\sqrt{k^2+m^2}t)}{\sqrt{k^2+m^2}}\hat{g}(k)$$

Thus $u(t)$ can be written

$$u(t) = \frac{\partial R}{\partial t} * f + R * g$$

where for each t, R is the inverse Fourier transform

$$R(x,t) = \frac{1}{\sqrt{2\pi}} \int_{-\infty}^{\infty} e^{ikx} \frac{\sin\sqrt{k^2+m^2}t}{\sqrt{k^2+m^2}} dk$$

of $(k^2 + m^2)^{-1/2}\sin\sqrt{k^2+m^2}t$ in the sense of distributions. The convolution $R * g$ makes sense since $R \in \mathcal{S}'(R)$ and $g \in \mathcal{S}(R)$. There are many ways to figure out what the function $R(x,t)$ is. Here is one. First we notice that $R(x,t)$ is zero except when $x^2 \leq t^2$. This follows from the Payley-Wiener theorem for distributions and the analyticity and growth of $(k^2 + m^2)^{-1/2}\sin\sqrt{k^2+m^2}\,t$ in the imaginary k directions. Second, we can compute directly that $R(x,t)$ is invariant under Lorentz transformations in two dimensions. Thus, $R(x,t)$ is a function of $t^2 - x^2$. Suppose that we write for $x^2 \leq t^2$,

$$H(\sqrt{t^2-x^2}) \equiv R(x,t) = \frac{1}{\sqrt{2\pi}} \int_{-\infty}^{\infty} e^{ikx} \frac{\sin\sqrt{k^2+m^2}\,t}{\sqrt{k^2+m^2}} dk$$

Then, differentiating twice with respect to t and twice with respect to x and subtracting the results, one finds that $H(\sqrt{t^2-x^2})$ satisfies:

$$m^2H''(\ \ t^2-x^2) + \frac{1}{t^2-x^2} H'(\ \ t^2-x^2) + H(\ \ t^2-x^2) = o$$

so

$$H(\ \ t^2-x^2) = cJ_o(m\ \ t^2-x^2)$$

where J_o is the first Bessel function. Setting $x = o$, one determines that $c = \frac{1}{2}$. Thus,

$$R(x,t) = \frac{1}{2} \chi_{\{x|x^2 \leq t^2\}}(x) \quad J_o(m\sqrt{t^2-x^2}) \tag{78}$$

Therefore, we have the representation:

$$(R * g)(x,t) = \frac{1}{2} \int_{-t}^{t} J_o(m\sqrt{t^2-y^2})\, g(x-y)\,dy \tag{79}$$

To analyse the decay of $R * g$ we need the estimates (see, for example, [23]):

$$J_o(\mu) = \left(\frac{2}{\mu\pi}\right)^{1/2} \cos(\mu - \frac{\pi}{4}) + O(\mu^{-3/2}) \tag{80}$$

$$J_1(\mu) = O(\mu^{-1/2})$$

as $\mu \longrightarrow \infty$. We write (79) as an integral over $\{y\,|\,|y| \leq \frac{t}{2}\}$ and an integral over $\{y|\frac{t}{2} \leq |y| \leq t\}$. Using $|J_o(\mu)| \leq c\mu^{-1/2}$, the first can easily be estimated by

$$ct^{-1/2} \int_{-t/2}^{t/2} |g(x-y)|\,dy \leq ct^{-1/2}\,||g||_1 \tag{81}$$

There are two integrals left one of which is

$$\left(\frac{1}{2\pi}\right)^{1/2} \int_{t/2}^{t} J_o(m\sqrt{t^2-y^2})\, g(x-y)\,dy$$

$$= \left(\frac{1}{2\pi}\right)^{1/2} \int_{t/2}^{t} \frac{\cos(m\sqrt{t^2-y^2}+\frac{\pi}{4})}{(t^2-y^2)^{1/4}}\, g(x-y)\,dy + \int_{t/2}^{t} O(\ (t^2-y^2)^{-3/4})\, g(x-y)\,dy$$

For the second term we have

$$\int_{t/2}^{t} O((t^2 - y^2)^{-3/4})g(x - y)dy$$

$$\leq ct^{-3/4}||g||_\infty \int_{t/2}^{t} (t - y)^{-3/4}dy$$

$$\leq ct^{-1/2}||g'||_1$$

To handle the first term, we integrate by parts obtaining:

$$- g(x-y)\ \frac{(t^2-y^2)^{1/4}}{my\ \sqrt{2\pi}} \sin(m\sqrt{t^2-y^2} + \frac{\pi}{4})\Bigg]_{y=t/2}^{y=t}$$

$$+ \frac{1}{m\sqrt{2}} \int_{t/2}^{t} \sin(m\sqrt{t^2-y^2} + \frac{\pi}{4})\ \frac{d}{dy}\left\{\frac{(t^2-y^2)^{1/4}}{y}\ g(x-y)\right\} dy$$

Both terms may be easily estimated by $ct^{-1/2}(||g||_1 + ||g'||_1)$. Combining this with (81) we have

$$||R*g||_\infty \leq ct^{-1/2}(||g||_1 + ||g'||_1)$$

To treat the $\frac{\partial R}{\partial t}*f$ term, notice that

$$\frac{\partial R}{\partial t} = \frac{1}{2}\ (\delta(x+t) + \delta(x-t)) + \frac{m}{2}\ \frac{t}{\sqrt{t^2-x^2}}\ J_1(m\sqrt{t^2-x^2})$$

so

(82) $$(\frac{\partial R}{\partial t}*f)(x,t) = \frac{1}{2}\left\{f(x+t) + f(x-t)\right\}$$

$$+ \frac{m}{2}\int_{-t}^{t} \frac{t}{\sqrt{t^2-y^2}}\ J_1(m\sqrt{t^2-y^2})f(x-y)dy$$

First we estimate the integral over $\{y|y^2 \leq t^2\}$ as before. Then we integrate the remaining integral by parts, observe that the boundary terms at $y = \pm t$ cancel the first term in (82), and estimate the other boundary terms and the remaining integral just as we did for $R*g$.

The result is

$$||(\frac{\partial R}{\partial t} * f)(x,t)||_\infty \le ct^{-1/2}(||f||_1 + ||f'||_1 + ||f''||_1)$$

The extra derivative on f occurs because there was one extra integration by parts. This proves the lemma. ∎

For f and g nice, this lemma gives us a decay estimate and leads us to define

$$||<u,v>||_a = ||u||_\infty$$

$$||<u,v>||_b = ||u||_1 + ||u'||_1 + ||u''||_1 + ||v||_1 + ||v'||_1$$

We take as our Hilbert space

$$\mathcal{H}_0 = \{\phi = <u,v> \,\big|\, ||\phi||^2 = ||Bu||_2^2 + ||v||_2^2 < \infty\}$$

Then, by (74), (i) holds. Notice that we don't quite have (ii) yet since we only know the estimate on $D = \mathcal{S}(R) \times \mathcal{S}(R)$. However, given $\phi \in \mathcal{H}_0$ with $||\phi||_b < \infty$, we can find a sequence $\phi_n \in D$ so that $||\phi - \phi_n||_b \longrightarrow 0$ and $||\phi - \phi_n|| \longrightarrow 0$. Since e^{-itA} is continuous and linear, it follows that

$$||e^{-itA}\phi||_\infty \le ct^{-1/2}||\phi||_b$$

since this holds for each n. Thus we have (ii). It remains to check for which p (iii) holds. Since $J(\phi) = <o,\lambda u^p>$,

$$\begin{aligned} ||J(\phi_1) - J(\phi_2)|| &\le |\lambda|\ ||u_1^p - u_2^p||_2 \\ &\le |\lambda|\ ||u_1 - u_2||_2(||u_1||_\infty + ||u_2||_\infty)^{p-1} \\ &\le c|\lambda|(||\phi_1||_a + ||\phi_2||_a)^{p-1}\ ||u_1 - u_2|| \end{aligned}$$

$$||J(\phi_1) - J(\phi_2)||_b = |\lambda|\{||u_1^p - u_2^p||_1 + ||D(u_1^p - u_2^p)||_1\}$$

$$\le c|\lambda|\{||u_1 - u_2||_2 + ||u_1' - u_2'||_2\}(||u_1||_2 + ||u_2||_2)(||u_1||_\infty + ||u_2||_\infty)^{p-2}$$

$$\leq \beta \, (||\phi_1||_a + ||\phi_2||_a)^{p-2} \, ||\phi_1 - \phi_2||$$

Thus, the estimates in hypothesis (iii) are satisfied with $q = p-2$. Since $d = \frac{1}{2}$ and $dq > 1$ we must choose $q > 2$ so $p > 4$. We have thus proven by Theorems 16, 17, and 18, global existence for small Cauchy data and the existence of the scattering operator for small data if $p > 4$. Notice that this result holds whatever the sign λ and whether p is either even or odd (or fractional).

In the case where p is odd and the sign of λ is negative, then we have global existence. Furthermore we can use the fact that the term

$$\int_{-\infty}^{\infty} u^{p+1} \, dx$$

in the conserved energy is bounded by $c||u||^{p+1}$ to show (like we did in Section 4) that the mapping $M_t : \phi(o) \longrightarrow \phi(t)$ is uniformly bounded on balls in $\mathcal{H}_o$. Thus, the hypotheses of Theorem 19 are satisfied so in these cases we have the existence of the wave operators. The completeness of the wave operators is an open question. We summarize:

<u>Theorem 20</u> (a) If $p > 4$, then global solutions of (75) exist if the initial data ϕ_o has small enough $||\phi_o||_{scat}$ norm (or the coupling constant λ is small enough). Further, the scattering operator exists and is continuous on such small data.

(b) If in addition p is odd and λ is negative the wave operators exist and are one to one maps (uniformly $||\cdot||_{scat}$ continuous on balls) of Σ_{scat} into itself.

In the proof of lemma 1 we used the explicit representation of the solution in terms of the Riemann function of the equation and its time derivative. This is the standard "old fashioned" technique (see [9]) There are other fancier ways to derive the $ct^{-1/2}$ decay of $||u||_\infty$. See for example [32], and [14]. However, if we want an explicit estimate for c in terms of a tractable norm on the initial data then it is easiest to use the old methods. The proof we give is the proof in Strauss [21] adapted to the case $n = 1$. A large number of decay estimates can be found in von Wahl [42], [43] and Costa [46].

Part C. $u_{tt} - \Delta u + m^2 u = \lambda u^p$, three dimensions

In order to handle the non-linear Klein-Gordon equation in three dimensions we first need a lemma which is analogous to lemma 1.

Lemma 2 Let $f, g \in C_o^\infty(R^3)$. And let $u(x,t)$ be the solution of

$$u_{tt} - \Delta u + m^2 u = o$$
$$u(x,o) = f(x)$$
$$u_t(x,o) = g(x)$$

Then, there is a universal constant c so that

$$||u(x,t)||_\infty \leq ct^{-3/2}||<f,g>||_b \tag{83}$$

where $||<f,g>||_b$ is defined as the sum of the L_1 norm of all the derivatives of f of order ≤ 3 and all the derivatives of g of order ≤ 2.

The proof of this lemma is similar to the proof of Lemma 1 (see Strauss [21]); only the determination of the form of $R(x,t)$ is a little more complicated. The extra derivative on the initial data comes about because $R(x,t)$ itself involves J_1. Thus, one must integrate by parts twice for the g terms and three times for the f term.

So, we choose $||\phi||_b$ to be as defined in the lemma and

$$||\phi||_a = ||u||_\infty$$

Then, (ii) is satisfied for nice data with $d = \frac{3}{2}$. However we can no longer use the Hilbert space $\mathcal{H}_o$ because it is not true that $||u||_\infty \leq c||Bu||_2$ in three dimensions. It is true that $||u||_\infty \leq c||B^2u||_2$ so we can use $\mathcal{H}_1$. That is, if we set

$$||<u,v>||^2 = ||B^2u||_2^2 + ||Bv||_2^2$$

$$\mathcal{H}_1 = \{<u,v> \mid ||<u,v>|| < \infty\}$$

then (i) holds. As in part B one can now extend (83) to all $\phi \in \mathcal{H}_1$ with $||\phi||_b < \infty$. Further, similar calculations to those in one dimension show that,

(84) $$||J(\phi_1) - J(\phi_2)|| = |\lambda| \; ||B(u_1^p - u_2^p)||_2$$

$$\leq |\lambda|(||\phi_1|| + ||\phi_2||)(||\phi_1||_a + ||\phi_2||_a)^{p-2}||\phi_1 - \phi_2||$$

There are many terms in $||J(\phi_1) - J(\phi_2)||_1$. Let us look at one of highest order in D (denote by D_i any partial derivative).

$$||D_i^2(u_1^p - u_2^p)||_1 \leq ||(D_i^2(u_1 - u_2))P(u_1,u_2)||_1$$

$$+ 2||(D_i(u_1 - u_2))D_iP(u_1,u_2)||_1$$

$$+ ||(u_1 - u_2)D_i^2P(u_1,u_2)||_1$$

$$\leq C(||Bu_1||_2 + ||Bu_2||_2)(||u_1||_\infty + ||u_2||_\infty)^{p-2} ||B^2(u_1 - u_2)||_2$$

$$+C(||Bu_1|| + ||Bu_2||_2)^2(||u_1||_\infty + ||u_2||_\infty)^{p-3} ||u_1 - u_2||_\infty \qquad (85)$$

$$\leq \beta \{(||\phi_1||_a + ||\phi_2||_a)^{p-3}||\phi_1 - \phi_2||_a + (||\phi_1||_a + ||\phi_2||_a)^{p-2} ||\phi_1 - \phi_2||$$

Thus, (iii) is satisfied with $q = p-2$ for slightly different reasons than in one dimension (you should not assume that $q = p-2$ is all right in all dimensions). Notice in the case $q = 1$, the constant β in (iii) is small if $||\phi_1|| + ||\phi_2||$ is small as required (see (84) and (85)). Since $d = \frac{3}{2}$ we need only choose $q \geq 1$, so $p \geq 3$. For all such p, Theorems 16 and 17 give a small data scattering theory for the equation

$$u_{tt} - \Delta u + m^2 u = \lambda u^p \quad , \quad x \in R^3 \qquad (86)$$

independent of whether p is even or odd and the sign of λ. Notice that the results are somewhat different in the two cases $p > 3$ and $p = 3$.

For $p \geq 3$ Theorem 18 gives global existence of strong solutions of the integral equation corresponding to (86). Furthermore, these solutions are strongly differentiable in t locally (see Section 2, part c). Thus we have global strong solutions of 86 for all $p \geq 3$ if the initial data is small enough (or λ is small enough). We also know the existence of global weak solutions for all data (Section 5). What remains open is the difficult problem of proving global strong solutions for large data.

Notice that as the dimension is increased (in our case from one to three) the scattering theory gets easier because the decay of free solutions is better (though $\| \ \|_b$ is a little more complicated) but the global existence theory gets harder because the Sobolev estimates are weaker. This puts us in the following peculiar position with regard to the wave operators. To use Theorem 19 to guarantee the existence of the wave operators we need to know global existence of strong solutions. In three dimensions we only know this for $p \leq 3$ (with p odd and λ negative). On the other hand the hypotheses of Theorem 19 exclude the borderline case $q = p-2 = 1$ and thus require $p > 3$. Therefore, as it stands, Theorem 19 says nothing about the existence of the wave operators in three dimensions if $p = 3$.

This provides us with an opportunity to illustrate a very important point. Namely, the hypotheses of the abstract theorems in these lectures are quite general in that they depend only on estimates relating A and J (and energy inequalities). Therefore, the conclusions are sometimes not as strong as those one can obtain by exploiting special properties of the equation being studied. Typically, one uses the same idea of proof as in Theorems 16, 17, 18, 19, but the step which fails on the general level is carried through using the special property.

To see how this works in the case $p = 3$ (where we know global existence) notice that the reason that Theorem 19 cannot handle the case $q = 1$ is that $(1+|t|)^d$ times the expression (63) on page 74 does not go to zero as $t \longrightarrow -\infty$ (because part (b) of the lemma excludes the case $q = 1$). Thus we need a new way to estimate:

$$\|e^{-iA(t-s)}J(\psi(s))\|_a = \|[B^{-1}\sin B(t-s)]u(s)^3\|_\infty$$

$$\leq C\|(k^2+m^2)^{-1/2}(\sin\sqrt{k^2+m^2}(t-s))\,\widehat{u^3}(k)\|_1$$

$$\leq C\|(k^2+m^2)^{-1}\|_2\|(k^2+m^2)^{1/2}\widehat{(u^3)}(k)\|_2$$

$$\leq C\|B(u(s)^3)\|_2$$

$$\leq C\|Bu(s)\|_2\,\|u(s)\|_\infty^2 .$$

Using this estimate in place of the ones before (63) the proof goes through as before. This how Segal originally proved the existence of the wave operators for $p = 3$ in three dimensions (see [32]). We summarize:

Theorem 21 ⓐ If $p \geq 3$, then global solutions of (86) exist if the initial data has small enough $||\phi_o||_{scat}$ -norm (or the coupling constant λ is small). Furthermore the scattering operator exists and is continuous on such small data.

ⓑ In the case $p = 3$ and λ negative, the wave operators exist.

Part D. The Coupled Dirac and Klein-Gordon Equations

In order to discuss the coupled Dirac and Klein-Gordon equations in three dimensions, we will first look at the non-linear Dirac equation itself. We can write the free Dirac equation as

$$\frac{\partial\psi}{\partial t}(x,t) = -i(i\vec{\alpha}\cdot\nabla + i\beta M)\psi(x,t) \tag{89}$$

$$= -iD_e\psi(x,t)$$

$$\psi(x,o) = f(x)$$

where

$$\psi(x,t) = \langle\psi_o(x,t),\ \psi_1(x,t),\ \psi_2(x,t),\ \psi_3(x,t)\rangle$$

$$f(x) = \langle f_o(x),\ f_1(x),\ f_2(x),\ f_2(x)\rangle$$

$$\vec{\alpha}\cdot\nabla = \alpha_1\frac{\partial}{\partial x} + \alpha_2\frac{\partial}{\partial y} + \alpha_3\frac{\partial}{\partial z}$$

The α_i's and β are certain anti-Hermetian 4x4 matrices. Thus, D_e is a formally self-adjoint operator on $\bigoplus_{i=0}^{3} L^3(R^3)$. If we let $B_e = \sqrt{-\Delta+M^2}$ on $L^2(R^3)$ then the anti-commutation properties of the α_i's and β (see [1] or [8]) imply that

$$D_e^2 = \begin{pmatrix} B_e^2 & & & 0 \\ & B_e^2 & & \\ & & B_e^2 & \\ 0 & & & B_e^2 \end{pmatrix} \equiv \vec{B}_e^2 \tag{88}$$

Further D_e is self-adjoint on the domain

$$D(D_e) = \bigoplus_{i=0}^{3} D(D_e)$$

and since D_e commutes with $\bar{B}_e^2$ it commutes with all powers of $\bar{B}_e$. Thus, we can take for our Hilbert space any of the escalated energy spaces:

$$\mathcal{K}_k = \{\psi \mid \|\psi\|_k^2 \equiv \sum_{i=0}^{3} \|B_e^k \psi_i\|_2^2 < \infty\}$$

Further, the group e^{-itD_e} generated by D_e takes each of these spaces into themselves and this implies ([26], p.269) that D_e is a self-adjoint operator on $\mathcal{K}_k$ with domain

$$D(D_e) = \bigoplus_{i=0}^{3} D(B_e^{k+1})$$

By the properties of the α's and β, each component $\psi_i(x,t)$ of $\psi(x,t)$ satisfies the Klein-Gordon equation

$$\frac{\partial^2}{\partial t^2}\psi_i + B_e^2\psi_i = 0$$

$$\psi_i(x,0) = f_i(x)$$

$$\frac{\partial \psi_i}{\partial t}(x,0) = g_i(x)$$

where g_i is the ith component of $-iD_e f$. Thus, by lemma 2 in part c,

$$\|\psi_i(x,t)\|_\infty \leq ct^{-3/2}\|f\|_b$$

where $\|f\|_b$ denotes the sum of the L^1 norms of all derivatives ≤ 3 of <u>all</u> the f_i's. Thus, if we define

$$\|\psi\|_a = \sum_{i=0}^{3} \|\psi_i(x,t)\|_\infty$$

we have

$$\|\psi\|_a \leq ct^{-3/2}\|f\|_b \tag{89}$$

We are now ready to treat the non-linear Dirac equation

$$\psi_t + iD_e\psi = \lambda J(\psi) \tag{90}$$

where we will assume that each component of $J(\psi)$ is a polynomial in the ψ_i's each of whose terms has order p. The application of Theorem 16, 17, and 18 differs in one respect from the application to the non-linear Klein-Gordon equation in part c. The non-linear term of part c, $J(\langle u,v\rangle) = \langle o,u^p\rangle$ was zero in the first term so the worst terms in $||J(\phi)||_b$ looked like $||B^2u^p||_1$. Here J looks like ψ^p (symbolically) in <u>each</u> component so the worst terms in $||J(\psi)||_b$ will look like $||B_e^3\psi^p||_1$. Thus, we cannot hope to satisfy the second part of hypothesis (iii) unless we choose as our Hilbert space $\mathcal{X}_1$ so that the norm

$$||\psi||^2 \equiv ||\psi||_3^2 \equiv \sum_{i=0}^{3} ||B_e^3\psi_i||_2^2$$

has three derivatives on each component. But, this means, for example, that $||J(\psi)||$ will be a sum of terms of the form $||B^3(\psi^p)||_2$. This term (and the other terms in hypothesis (iii)) can be estimated just as in part c but now we will have $q = p-3$ (instead of $p - 2$) so we must require $p \geq 4$ in order to apply Theorem 16, 17, and 18.

<u>Theorem 22</u> Suppose that each component of $J(\psi)$ is a polynomial in the ψ_i each term of which has degree at least 4. Then the non-linear Dirac equation (90) in three dimensions has global strong solution if the Cauchy data is sufficiently small (in the $||\cdot||_{scat}$ norm). Furthermore on these small data the scattering operator exists and is continuous.

Now we can set up the spaces and norms for the coupled Dirac-Klein-Gordon equations:

$$\begin{aligned} \frac{d}{dt}\psi + iD_e\psi &= \lambda_e J_e(\psi,u) \\ u_{tt} - \Delta u + m_o^2 u &= \lambda_o F_o(\psi,u) \end{aligned} \tag{91}$$

which we rewrite as

$$\begin{aligned} \psi_t + iD_e\psi &= \lambda_e J_e(\psi,\phi) \\ \phi_t + iA_o\phi &= \lambda_o J_o(\psi,\phi) \end{aligned} \tag{92}$$

where $\phi = \langle u,v\rangle$, $B_o = \sqrt{-\Delta + m_o^2}$, $J_o(\psi,\phi) = \langle 0,F_o(\psi,u)\rangle$, $J_e(\psi,\phi) = J_e(\psi,u)$, and

$$A_o = i\begin{pmatrix} o & I \\ -B_o^2 & o \end{pmatrix}$$

as usual. We take as our Hilbert space

$$\mathcal{H} = \mathcal{K}_3 \oplus \mathcal{H}_3 = (\bigoplus_{i=0}^{3} D(B^3)) \oplus (D(B^3) \oplus D(B^2))$$

with the norm on $\Xi = \langle\psi,\phi\rangle$ given by

$$||\Xi||^2 = \sum_{i=0}^{3} ||B^3\psi_i||_2^2 \quad + \quad ||B^3 u||_2^2 + ||B^2 u||_2^2$$

Then the operator

$$A = \begin{pmatrix} D_e & 0 \\ 0 & A_0 \end{pmatrix}$$

is self-adjoint on

$$D(A) = (\bigoplus_{i=0}^{3} D(B^4)) \oplus (D(B^4) \oplus D(B^3))$$

and if we set $J(\Xi) = \langle\lambda_e J_e(\Xi), \lambda_0 J_0(\Xi)\rangle$ then we can write (92) as

$$\Xi'(t) = -iA\Xi(t) + J(\Xi(t)) \tag{93}$$

which is in the right form to apply the theorems of these lectures. We define $||\Xi||_b$ to be the sum of the L^1 norms of all the derivatives of ψ_i, $i=0,1,2,3$, and u of order ≤ 3 plus the L^1 norms of all derivatives of v of order ≤ 2. And, we set

$$||\Xi||_a = \sum_{i=0}^{3} ||\psi_i||_\infty + ||u||_\infty$$

The three norms $||\cdot||$, $||\cdot||_a$, $||\cdot||_b$ satisfy hypotheses (i) and (ii) with $d = 3/2$, so it only remains to check what properties J_e and J_0 must have so that (iii) is satisfied. From the calculations on the non-linear Dirac equation and part c you should be able to guess the answer. The proofs are exactly like the calculations we have already done. So, from Theorems 16, 17, 18, we have:

<u>Theorem 23</u> Suppose that $F_0(\psi,u)$ and each component of $J_e(\psi,u)$ are polynamials in the ψ_i and u so that each term in the components of J_e has order at least 4 and each term in F_0 has order at least 3. Then the coupled equations (91) have unique global solutions if the $||\cdot||_{scat}$

norms of the initial data are small enough (or for any fixed initial data in Σ_{scat} if λ_o and λ_e are taken small enough). Furthermore the scattering operator exists on these small initial data and is continuous.

This theorem does not cover the case of a Yukawa interaction

$$J_e(\Xi) = \lambda_e u\bar{\psi}\gamma_o\psi \quad , \quad J_o(\Xi) = \lambda_o\bar{\psi}\gamma_o\psi$$

because the degrees are too small, but it does cover generalized Yukawa interactions of the form

$$J_e(\Xi) = \lambda_e u(\bar{\psi}\gamma_o\psi)^k \quad , \quad J_o(\Xi) = \lambda_o(\bar{\psi}\gamma_o\psi)^k$$

for any $k \geq 2$. You can easily formulate for yourself the analogous theorem for two coupled Fermions

$$\frac{d}{dt}\psi^{(1)} - iD_e^{(1)}\psi^{(1)} = \lambda_e^{(1)} J_e^{(1)}(\psi^{(1)},\psi^{(2)})$$

$$\frac{d}{dt}\psi^{(2)} - iD_e^{(2)}\psi^{(2)} = \lambda_e^{(2)} J_e^{(2)}(\psi^{(1)},\psi^{(2)})$$

where degree four is required in both terms. Chadam applied these small data ideas to the classical versions of the equations of quantum field theory in [6]. He uses more delicate (L^p) decay estimates and more special properties of the Dirac and Klein-Gordon propagators so he can handle more cases. For example, in the Fermion-Fermion case above, he only requires degree 3.

12. Asymptotic Completeness

We come now to a really hard question, the problem of asymptotic completeness. Let us suppose that the hypotheses of Theorem 19 are satisfied so that the wave operators $\Omega_\pm$ exist and are continuous one to one maps of Σ_{scat} into itself. If $\Omega_\pm$ are asymptotically complete, that is, if

$$\text{Range } \Omega_+ = \text{Range } \Omega_-$$

then we can define the scattering operator

$$S = (\Omega_+)^{-1}\Omega_-$$

To see what is involved let ϕ_o be in the range of Ω_-. Thus there is a $\phi_- \in \Sigma_{scat}$ and a solution of

$$\phi(t) = e^{-iAt}\phi_- + \int_{-\infty}^{t} e^{-iA(t-s)}J(\phi(s))ds$$

so that

$$\phi(o) = \phi_o$$

and

$$||\phi(t) - e^{-itA}\phi_-|| \longrightarrow o \quad \text{as} \quad t \longrightarrow -\infty$$

What we must prove is that there is a $\phi_+ \in \Sigma_{scat}$ so that

(94) $$||\phi(t) - e^{-itA}\phi_+|| \longrightarrow o \quad \text{as} \quad t \longrightarrow +\infty$$

If (94) is true then

$$\phi_+ = \lim_{t\to\infty} e^{itA}\phi(t)$$

since e^{-itA} is unitary, so the natural way to construct ϕ_+ is to show that $e^{iAt}\phi(t)$ is a Cauchy sequence in $\mathcal{H}$. By hypotheses (iii) we have

$$||e^{it_1A}\phi(t_1) - e^{it_2A}\phi(t_2)|| \leq \int_{t_2}^{t_1} ||J(\phi(s))||ds$$

$$\leq \beta \int_{t_2}^{t_1} ||\phi(s)||_a^q ||\phi(s)||ds$$

In general we get an upper bound on $||\phi(s)||$ from an energy inequality, so what we need to know, to insure that the right hand side goes to zero as $t_1, t_2 \longrightarrow +\infty$, is that

$$\int_{-\infty}^{\infty} ||\phi(s)||_a^q \, ds < \infty$$

For example, in the case of the equation

$$u_{tt} - \Delta u + m^2 u = -u^3 \qquad x \in R^3 \tag{95}$$

the required estimate is

$$\int_{-\infty}^{\infty} ||u(x,t)||_\infty^2 \, dt < \infty$$

What is needed therefore is an apriori estimate on solutions of the non-linear equation which guarantees that solutions with nice initial data decay sufficiently rapidly in the $||\cdot||_a$ norm as $t \longrightarrow \pm\infty$. Furthermore, if we expect to be able to prove continuity of the scattering operator we need to estimate the constants in this decay in terms of the decay of the corresponding solution of the free equation. For most non-linear wave equations no such apriori estimates are known (or only very weak estimates) so one cannot even begin to attack the problem of asymptotic completeness. However, in the case of (95) in three dimensions this problem has been solved by Morawetz and Strauss [21]. Their theorem states:

Theorem 21 Let $\mathfrak{F}$ denote the closure of $C_0^\infty(R^3) \times C_0^\infty(R^3)$ in the norm

$$||\langle f,g\rangle||_{scat}^2 = \sup_t \left\{\int_{R^3} (u_t^2 + |\nabla u|^2 + m^2 u^2)\,dx\right\}$$

$$+ \sup_t \sup_x |u(x,t)|^2 + \int_{-\infty}^{\infty} \sup_x |u(x,t)|^2 dt$$

where $u(x,t)$ denotes the solution of the free equation $u_{tt} - \Delta u + m^2 u = 0$ with initial data $u(x,0) = f, u_t(x,0) = g$. Then the scattering operator for (95) exists and is a continuous one to one map of $\mathfrak{F}$ onto itself Further, if $\langle f,g\rangle \in \mathfrak{F}$ and ∇f has finite energy, the solution $\tilde{u}(x,t)$ of (95) with initial data $\langle f,g\rangle$ satisfies

$$||\tilde{u}(x,t)||_\infty \leq C(1+|t|)^{-3/2}$$

The crucial fact which made this theorem possible was a new apriori estimate which was derived from a weak apriori estimate previously obtained by Morawetz [20]. I recommend this argument to anyone who wants to see how beautiful and difficult these non-linear theories are (just in case you are not already convinced). The same arguments go through for a fairly restricted class of other non-linear terms described in an appendix in [21]. Morawetz and Strauss prove further properties of S in [22].

Basically, there are two kinds of arguments in [21]. The arguments used to derive the apriori estimate and other necessary estimates use the explicit representation of the Riemann function for the free equation and methods from partial differential equations. These techniques are necessarily directly related to the particular differential equation being studied and therefore do not generalize in a natural way to the abstract level. However, the second part of the Morawetz-Strauss argument is essentially a functional analysis argument which shows that, given the estimates, one can construct the scattering operator. This part of the argument can be formulated on the abstract level and it is the purpose of the rest of this section to outline how this may be done. Since the most important thing is the general structure of the argument we will just outline the ideas. Details may be found in Reed [25].

Since there are quite a few hypotheses, we will collect them all here rather than refer back to previous sections. The hypotheses fall naturally into four parts:

I (Existence) Let A be a self-adjoint operator on a Hilbert space $\mathcal{H}$ (with norm $||\cdot||$) and let J be a non-linear mapping on $\mathcal{H}$ which satisfies

(96) $$||J(\phi) - J(\psi)|| \leq C(||\phi||,||\psi||)\ ||\phi - \psi||$$

Then, for each $\phi_0 \in \mathcal{H}$, the corollary of Theorem 1 gives a local solution of

(97) $$\phi(t) = e^{-iAt}\phi_0 + \int_0^t e^{-iA(t-s)} J(\phi(s))ds$$

We denote by M_t the map $M_t : \phi(o) \to \phi(t)$ and assume that $||M_t\phi_o||$ is apriori bounded (typically the result of an energy inequality) which implies that the solution $\phi(t) = M_t\phi_o$ is global in t. Finally, we assume that there is a dense set D in $\mathcal{H}$ so that $e^{-iAt} : D \longrightarrow D$ and $M_t : D \longrightarrow D$. In applications D would typically be of the form $\bigcap_{n=1}^{\infty} D(A^n)$ or equal to the set of C^∞ Cauchy data with compact support.

II (Decay of free solutions) We assume that there is an auxiliary norm $||\cdot||_a$ on D so that

$$(98)\quad ||J(\phi) - J(\psi)|| \leq C(||\phi||,||\psi||)\ ||\phi - \psi||\ (||\phi||_a^r + ||\psi||_a^r)$$

for some $r > o$ and all $\phi,\psi \in D$. If $\phi(\cdot)$ is a continuous (in the $||\cdot||$ norm) D-valued function on t_1,t_2, we define,

$$|||\phi(\cdot)|||^r_{t_1,t_2} = \sup_{t_1\leq t\leq t_2} ||\phi(t)||^r + \sup_{t_1\leq t\leq t_2} ||\phi(t)||_a^r + \int_{t_1}^{t_2} ||\phi(s)||_a^r\, ds$$

When $t_1 = -\infty$ and $t_2 = \infty$, we denote the norm simply by $|||\cdot|||$. If $\phi \in D$, we assume that

$$(99)\qquad |||e^{-itA}\phi||| < \infty$$

and define

$$||\phi||_{scat} = |||e^{-itA}\phi||| \qquad \phi \in D$$

The Banach space of scattering states Σ_{scat} is by definition the closure of D in the $||\cdot||_{scat}$ norm.

III (the apriori estimate) We assume that for all $\phi \in D$,

$$(100)\qquad |||M_t\phi||| \leq f(||\phi||_{scat})$$

where f is a locally bounded function on (o,∞).

IV. (Kernel Estimates) Let $\phi(\cdot)$ and $\psi(\cdot)$ be continuous D-valued functions on $[t_1, t_2]$. Define for each pair (t_1, t_2),

$$[[\phi(\cdot)]]^r_{t_1,t_2} = \int_{t_1}^{t_2} ||\phi(t)||_a^r\, ds$$

and

$$J_{t_1,t_2}(\phi)\ (t) = \int_{t_1}^{t_2} e^{-iA(t-s)} \quad J(\phi(s))ds$$

We assume that J_{t_1,t_2} satisfies:

$$|||J_{t_1,t_2}(\phi) - J_{t_1,t_2}(\psi)||| \le C(|||\phi(\cdot)|||_{t_1,t_2}, |||\psi(\cdot)|||_{t_1,t_2})$$

(101) $$\cdot([[\phi]]_{t_1,t_2} + [[\psi]]_{t_1,t_2})^{\alpha_1} |||\phi(t) - \psi(t)|||_{t_2}^{\alpha_2}$$

$$\cdot(\sup_{t_1 \le t \le t_2} ||\phi(t) - \psi(t)||^{\alpha_3})$$

and

(102) $$|||J_{t_1,t_2}(\phi)||| \le C(|||\phi(\cdot)|||_{t_1,t_2})[[\phi]]_{t_1,t_2}^{1+\alpha_4}$$

where the α_i (which are independent of $\phi(\cdot)$ and $\psi(\cdot)$) satisfy $\alpha_i > 0$, $\alpha_1 + \alpha_3 \ge 1$, $\alpha_2 + \alpha_3 \ge 1$. (101) and (102) are further assumed to hold in the case where $t_2 = t$ on the left and $t_2 = \infty$ on the right and similarly for $t_1 = t$ and $t_1 = -\infty$.

If we let X_{t_1,t_2} denote the closure of D in the $|||\ |||_{t_1,t_2}$ norm, then what we are essentially assuming is that J_{t_1,t_2} is locally Hölder continuous, in both the $|||\ |||_{t_1,t_2}$ and the $\sup_{t_1 \le t \le t_2} ||\ ||$ norms, from X_{t_1,t_2} to $X_{-\infty,\infty}$, where the constant depends appropriately on $|||\phi|||_{t_1,t_2}$ and $|||\psi|||_{t_1,t_2}$. Since $\sup_{t_1 \le t \le t_1} ||\phi(\ )|| \le |||\phi(\ )|||_{t_1,t_2}$ and $\alpha_2 + \alpha_3 \ge 1$, J_{t_1,t_2} is actually Lipschitz in the $|||\ |||_{t_1,t_2}$ norm. In applications, such estimates are typically proven by expressing $e^{-iA(t-s)}$ as an integral operator and using L^p estimates on the kernel and non-linear term J. A simple example of such an estimate appears in Section 11, part c. For other examples, see Segal[32],[32], Chadam [4],[6], Morawetz and Stauss[21], and Strauss[35].

The proof of the existence of the scattering operator begins by investigating the properties of $\{M_t\}$ and $\{e^{-iAt}\}$.

Lemma 1 The families of mappings, $\{e^{-itA}\}$, $\{M_t\}$, and $\{e^{itA}M_t\}$, are uniformly equicontinuous on $||\cdot||_{scat}$-balls in D.

The proof of Lemma 1 proceeds by first showing that M_t is $||\cdot||$-continuous by using the kernel estimates and iteration. Then this is used to show that $\{e^{itA}M_t\}$ is an equicontinuous family similarly to the proof of continuity in Theorem 17. Since e^{-itA} is, for each t, linear, bounded and of norm one on Σ_{scat} the lemma follows. Lemma 1 shows in particular that for each t, M_t extends uniquely to a uniformly continuous map of Σ_{scat} into itself. Since the extension satisfies the integral equation (97) it coincides with the restriction of M_t on $\mathcal{H}$ to Σ_{scat}.

The following lemma shows that the apriori estimate also extends to Σ_{scat}.

Lemma 2 There exists a locally bounded function $\tilde{f}$ on $(0,\infty)$ so that $|||M_t\phi||| \leq \tilde{f}(||\phi||_{scat})$ for all $\phi \in \Sigma_{scat}$.

Lemma 2 is proven by choosing a sequence $\phi_n \in D$ so that $||\phi_n - \phi||_{scat} \longrightarrow 0$. First one shows that $M_t\phi_n$ converges to $M_t\phi$ and then one takes the limit in the apriori estimate which gives lemma 2 with $\tilde{f}(x) = \overline{\lim_{y\to x}} f(y)$

We can now prove the existence of asymptotic states:

Theorem 25 For each $\phi \in \Sigma_{scat}$ there exist ϕ_+ and ϕ_- in Σ_{scat} so that

$$||M_t\phi - e^{-itA}\phi_+|| \longrightarrow 0 \quad \text{as} \quad t \longrightarrow +\infty$$

$$||M_t\phi - e^{-itA}\phi_-|| \longrightarrow 0 \quad \text{as} \quad t \longrightarrow -\infty$$

The maps $\phi \xrightarrow{\Omega_+^{-1}} \phi_+$, $\phi \xrightarrow{\Omega_-^{-1}} \phi_-$ are continuous and one to one from Σ_{scat} into Σ_{scat}.

Proof We will give the first part of the proof since it shows how the apriori estimate and the kernel estimates are used. Let $\phi \in \Sigma_{scat}$ be given and fix t_1 and t_2. Then,

$$||e^{it_1A}M_{t_1}\phi - e^{it_2A}M_{t_2}\phi||_{scat} = |||e^{-iuA}(e^{it_1A}M_{t_1}\phi - e^{it_2A}M_{t_2}\phi)|||$$

$$= |||\int_{t_1}^{t_2} e^{-iA(u-s)} J(M_s\phi)ds|||$$

$$\leq |||[J_{t_1,t_2}(M_s\phi)](u)|||$$

$$\leq C(|||M_s\phi|||_{t_1,t_2})\ [[M_s\phi]]_{t_1,t_2}^{1+\alpha_4}$$

$$\leq C(||\phi||_{scat})\ [[M_s\phi]]_{t_1,t_2}^{1+\alpha_4}$$

By the apriori estimate we know that $|||M_t\phi||| < \infty$ which implies that

$$[[M_s\phi]]_{t_1,t_2}^{1+\alpha_4} \longrightarrow o \quad \text{as } t_1,t_2 \longrightarrow \infty$$

Thus $e^{itA}M_t\phi$ is Cauchy in $\sum_{scat}$ as $t \longrightarrow +\infty$ and the same is true as $t \longrightarrow -\infty$, so we can define

$$\phi_+ = \lim_{t\to+\infty} e^{itA}M_t\phi \quad , \quad \phi_- = \lim_{t\to-\infty} e^{itA}M_t\phi$$

The convergence statements in the theorem follow from the unitarity of e^{-itA}. By lemma 1, the maps Ω_+^{-1} and Ω_-^{-1} are pointwise limits of uniformly equicontinuous families of maps and are thus continuous. The proof that $\Omega_{\pm}^{-1}$ are one to one uses the kernel estimates and same trick as in the one to one proof in Theorem 16.

In order to construct the scattering operator we need to know that the maps $\Omega_{\pm}^{-1}$ just constructed take $\sum_{scat}$ onto $\sum_{scat}$ and that $\Omega_{\pm}$ are continuous. To prove this we must solve the Cauchy problem at $\pm\infty$. Given $\phi_+ \in \sum_{scat}$, the solution of (97) which satisfies

$$||\phi(t) - e^{-iAt}\phi_+|| \longrightarrow o \quad \text{as } t \longrightarrow +\infty$$

should be the solution of

(103) $$\phi(t) = e^{-itA}\phi_+ - \int_t^\infty e^{-iA(t-s)} J(\phi(s))ds$$

There are two difficulties. First we must show that (103) has a solution with the right decay properties, that is,

$$|||\phi(t)||| < \infty , \quad |||e^{-itA}\phi(o)||| < \infty$$

But, this is not quite enough. To conclude that $\phi(o) \in \Sigma_{scat}$ we must exhibit it as a limit of vectors in D since Σ_{scat} was defined as the closure of D in the $||\cdot||_{scat}$ norm. Thus, we define $\phi^N(t)$ to be the solution of

$$\text{(104)} \qquad \phi^N(t) = e^{-iAt}\phi_+ - \int_t^N e^{-iA(t-s)} J(\phi(s))ds$$

Since $\phi^N(N) = e^{-iAN}\phi_+$, $\phi^N(t)$ is just the unique global solution of (97) with Cauchy data $e^{-iNA}\phi_+$ at $t = N$ guaranteed by the hypotheses in I. In particular, if $\phi_+ \in D$, then $\phi^N(t) \in D$ for all t. What must be proven is that as $N \longrightarrow \infty$, $\phi^N(t)$ converges (pointwise in Σ_{scat}) to a solution of (103).

<u>Lemma 3</u> Let $\phi_+ \in D$ and let $\phi^N(t)$ be the corresponding solution of (104). Then, if T is large enough,

$$\text{(a)} \qquad |||\phi^N(t)|||_{T,\infty} \leq 2|||e^{-itA}\phi_+|||_{T,\infty}$$

$$\text{(b)} \qquad [[\phi^N(t)]]_{T,\infty} \leq 2[[e^{-itA}]]_{T,\infty}$$

The point of this lemma is that the right hand sides are independent of N and thus give us some control of the limit of the $\phi^N(t)$ as $N \to \infty$. Lemma 3 is proven by defining the space B(T) to be the set of $\phi(\cdot)$ in $X_{T,\infty}$ so that

$$|||\phi(t) - e^{-itA}\phi_+|||_{T,\infty} \leq |||e^{-itA}\phi_+|||_{T,\infty}$$

$$[[\phi(t) - e^{-itA}\phi_+]]_{T,\infty} \leq [[e^{-itA}\phi_+]]_{T,\infty}$$

and then showing that for T large enough the solution $\phi^N(t)$ of (104) lies in B(T) for all $N > T$. The proof uses the strict positivity of the α_i. Next we have,

Lemma 4 Let $\phi^N(t)$ be the solutions discussed above. Then,

(a) as $N \to \infty$, $\phi^N(\cdot)$ converges in $X_{T,\infty}$ to a function $\phi(t)$ which satisfies (103).

(b) $|||\phi(t)|||_{T,\infty} \leq 2|||e^{-itA}\phi_+|||_{T,\infty}$

(c) $\phi(t) \in \sum_{scat}$ for each $t \in [T,\infty)$ and $|||e^{itA}\phi(t) - \phi_+|||_{scat} \to 0$ as $t \to +\infty$.

To prove lemma 4 one first uses the kernel estimates and the uniformity in Lemma 3 to show that $\phi^N(\cdot)$ is Cauchy in $X_{T,\infty}$. It is easy to check that the pointwise limit $\phi(t)$ satisfies (103) and the estimate in (b) follows from the uniform estimates on the ϕ^N. The proof of the statements in (c) also use the uniform estimates from Lemma 3.

For fixed $\phi_+ \in D$ we can now define

$$\Omega_+^T : \phi_+ \longrightarrow \phi(0)$$

$$\Omega_+ : \phi_+ \longrightarrow M_{-T}\Omega_+^T\phi_+$$

Ω_+ is thus a map from D into $\sum_{scat}$ and (by Theorem 24 and its proof) $\Omega_+^{-1}\Omega_+ \phi_+ = \phi_+$. Similar definitions and statements hold for Ω_-^T, Ω_-. What remains to be shown is that Ω_+ can be extended to all of $\sum_{scat}$ and that the extension is continuous.

Theorem 26 (a) Let $\phi_+ \in \sum_{scat}$; then there is a T and a $\sum_{scat}$-valued function $\phi(\cdot)$ which satisfies (103) and parts (b) and (c) of Lemma 4.

(b) The map $\Omega_+ : \phi_+ \longrightarrow M_{-T}\Omega_+^T\phi_+$ is a continuous map of $\sum_{scat}$ into $\sum_{scat}$.

If Ω_+ were uniformly continuous on $||\cdot||_{scat}$ balls in D then the proof of Theorem 26 would be easy; we would just extend Ω_+ directly from D. But, this is not at all obvious since the choice of T depends not only on $||\phi_+||_{scat}$ but also on $[[e^{-itA}\phi]]_{T,\infty}$. Thus, one needs some local uniformity on T. First one chooses a sequence $\phi_+^n \in D$ so that $||\phi_+^n - \phi||_{scat} \longrightarrow 0$. Let $\phi^n(t)$ be the corresponding solutions of (103). Then one first shows that there is an N so that $n \geq N$ implies that there is a T so that

$$|||e^{-itA}\phi_+^n||| \leq 2k \quad , \quad [[e^{-itA}\phi_+^n]]_{T,\infty} \leq \theta_0$$

for a constant k and a small constant θ_0.
This uniformity and the fact that $\phi^n(t)$ satisfies (103) allow one to show that $\phi^n(\cdot)$ are Cauchy in $X_{T,\infty}$ and converge to a solution $\phi(t)$ of (103) which has the right properties.

To prove continuity one uses the same idea to choose a small η_0 so that for all ψ_+ in the ball

$$B(\eta_0,\phi_+) = \{\psi_+ \in \textstyle\sum_{scat} \big| \; ||\phi_+ - \psi_+||_{scat} \leq \eta_0\}$$

one can use the same T and uniform estimates

$$|||\psi(t)|||_{T,\infty} \leq 2||\phi_+||_{scat}$$

$$[[\psi(t)]]_{T,\infty} \leq 2\,[[e^{-itA}\phi_+]]_{T,\infty}$$

hold. With these uniform estimates the proof of continuity proceeds by iteration much like the proof of Theorem 17 to show that Ω_+^T is uniformly continuous on $B(\eta_0,\phi_+)$. From this one concludes that $\Omega_+ = M_{-T}\Omega_+^T$ is uniformly continuous on $B(\eta_0,\phi_+)$ since M_t is uniformly continuous on $\sum_{scat}$ balls for each t.

Of course, analogous statements to those on Theorem 25 hold for Ω_-. Thus, combining Theorem 25 and Theorem 26 we have:

Theorem 27 The scattering operator $S = \Omega_+^{-1}\,\Omega_-$ is a one to one continuous map of $\sum_{scat}$ onto $\sum_{scat}$.

I have two remarks to make about this theorem. First, although it looks very nice, remember that we assumed the existence of an apriori estimate (hypothesis III). If the experience of Morawetz and Strauss is any guide, the proof of such an apriori estimate is the hardest part of the problem. Secondly, there is no reason to think that our choice of norms, spaces, and estimates in hypotheses II and IV give the best general abstract result or are the only choices possible. In fact, in a particular application, one may have an apriori estimate but some of the hypotheses in II or IV may fail so that one can't apply Theorem

27 directly. In such a case one will have to choose other norms and spaces which are better suited to the problem at hand. Some technical details of the construction of the scattering operator will then be different but the general idea should follow the outline presented here.

13. Discussion

It should be clear from the preceeding sections that the scattering theory for non-linear wave equations consists mostly of unsolved or partially solved problems. Nevertheless, it is worthwhile to point out and discuss some of these problems explicitly. For convenience, we will group them into four parts, progressing from easier to harder problems.

First of all, there are many equations in the physics and engineering literature to which Theorem 16, 17, 18 or variants of them can be applied. Such applications would be relatively straightforward but not trivial since one must prove decay estimates for the free equation and choose the norms $\| \ \|, \| \ \|_a, \| \ \|_b$ correctly. Such work would not advance the mathematical theory very much, but would provide greater understanding of equations of direct physical interest.

The second general problem is to take the small data scattering operators or the wave operators that are known to exist and to investigate their properties. For example, do they commute with the natural representation of the Lorentz group (or other symmetry groups) on Σ_{scat} (see [22])? Another interesting question would be to take more complicated but tractable non-linear terms and investigate the behavior of the wave operators as functions of various parameters of the theory. For example, consider the coupled equations

$$(u_1)_{tt} - \Delta u_1 + m_1^2 u_1 = -4\lambda(u_1 + \alpha u_2)^3 - 2\beta u_1 u_2^2$$

$$(u_2)_{tt} - \Delta u_2 + m_2^2 u_2 = -4\lambda\alpha(u_1 + \alpha u_2)^3 - 2\beta u_2 u_1^2$$

These equations have a positive conserved energy if $\lambda > 0$, $\beta > 0$ and α is any real number and using the techniques in Section 11, part c the wave operators can be shown to exist. It should be interesting to investigate the analyticity properties of Ω_+ as functions of λ, β, α. One can easily imagine in the case of non-linear coupled Dirac equations right hand sides which display various internal symmetries. It would be nice to prove that the small data scattering operators have the same symmetries or to exhibit a case where the symmetry is broken.

Another interesting problem is to prove that the scattering operator has the form S = I + T where T is a "small" non-linear operator. Intuitively, this must be true for the small data scattering operator but it should be proven. Whether such a decomposition is true for the wave operators given by Theorem 19 is another more difficult question.

Ideally one would like to show that S (or $\Omega_{\pm}$) can be expanded as

$$S = I + \sum_{n=1}^{\infty} \lambda^n T_n$$

where λ is, for example, a small coupling constant and the T_n are (at least for low n) simple operators. This would allow one to calculate the scattering operator approximately. Such expansions or representations are well-known in linear theories (for example, see [28] in the quantum mechanical case and [17] for the case of classical <u>linear</u> wave equations). It is clear that we could go on and on with this list of questions about S and $\Omega_{\pm}$ but the above examples give the idea. Theorems 17 and 19 guarantee the existence of certain non-linear operators. The problem is to investigate the properties of those non-linear operators and how the properties reflect the structure of the non-linearities in the original equation.

The third general problem is to develop new techniques for handling the small data scattering theory and the existence of the wave operators when the non-linearity is not sufficiently high or the decay is too slow to allow application of the techniques we have presented. For example, consider the equation

(105) $$u_{tt} - u_{xx} + m^2 u = -u^3$$

in one-dimension. In order to prove the existence of the wave operators by the techniques we have outlined, one must have that (see Section 11, part c),

$$\int_T^\infty ||e^{-iA(t-s)} J(\phi(s))||_a ds = \int_T^\infty ||B(u(s)^3)||_2 ds$$

$$\leq \int_T^\infty ||Bu(s)||_2 \, ||u(s)||_\infty^2 ds$$

$$\longrightarrow 0 \quad \text{as} \quad T \longrightarrow \infty.$$

The term $||Bu(s)||_2$ is of course bounded by the energy, but in one dimension free solutions $u(s)$ only decay like $s^{-1/2}$ in the sup norm so we can't expect this convergence to hold. Nevertheless, it is clear that there should be a scattering theory for (105) in terms of solutions of the linear equation

(106) $$u_{tt} - u_{xx} + m^2u = 0$$

The rate of convergence of solutions of (105) to the appropriate solution of (106) will be slower; so slow in fact that one may have to use other norms besides the energy norm. We have picked an illustration where the divergence of (105) is borderline. There are lots of other cases where the divergence of integrals in the methods of Theorems 17 and 19 is much faster, but where a scattering theoty should nevertheless exist. This third problem area is much harder than the first two because it really requires techniques which go beyond those described in these lectures. The best approach is, I think, to concentrate on a particular example, do whatever is necessary to get a scattering theory there, and then see what this suggests about new more general techniques.

The fourth problem is, of course, the question of asymptotic completeness. Because of the necessity of deriving apriori decay estimates on solutions of non-linear equations, this problem is very difficult, and as we have mentioned before, complete results are known only in one class of examples. Any progress on this problem for any equations not handled by Morawetz and Strauss[21] would be very important. There is one aspect of this problem about which there seems to be some confusion. A soliton is a solution of a non-linear equation which keeps its form, for example a solution of the form $u(x+t)$. Soliton solutions of non-linear wave equations have generated much interest among physicists lately and there is a growing literature of applications (see[29]). My point that I want to emphasize here is that the presence of soliton solutions does not mean that there is no scattering theory. There are two cases to consider. First, suppose that the soliton solutions are not "normalized", that is they are not small enough as $x \longrightarrow \infty$ to be in the Hilbert spaces of the problem (many of the soliton solutions approach non-zero constants at $+\infty$). In this case there is no apriori reason to think that the presence of the soliton solutions should affect the scattering theory for data which is small at $x = \pm\infty$, which is the case most naturally handled by the Hilbert space methods we have discussed. Essentially, the soliton solutions should play no role because they are not in the class of initial data under discussion.

The more interesting case is where the solitons solutions are in the Hilbert space under discussion either because they are small enough at $x = \pm\infty$ or because we choose our Hilbert space norm so that initial data which are large at infinity are allowed. If ϕ_0 is the initial data for such a soliton solution, then we would not expect the soliton $M_t\phi_o$ to decay into free equations at $t = \pm\infty$ since the wave keeps its shape. But this does not preclude a complete scattering theory, it just says that we should expect that Range Ω_+ and Range Ω_- will be strictly contained in $\sum_{scat}$. If one has asymptotic completeness,

$$\text{Range } \Omega_- = \text{Range } \Omega_+$$

then one has the scattering operator $S = \Omega_+^{-1}\Omega_-$. Given a $\phi_- \in \sum_{scat}$, and setting $\phi_+ = S\phi_-$, then, if we send in a free wave $e^{-iAt}\phi_-$ in the distant past, we will get out a free wave $e^{-iAt}\phi_+$ in the distant future. This situation is similar to the situation in quantum mechanics where one expects that the ranges of the wave operators equal the part of the Hilbert space corresponding to the absolutely continuous part of the spectrum of the interaction Hamiltonian H_I. In general, H_I will have bound states which will not decay to free solutions but this does not prevent the construction of a scattering theory. Of course, in the quantum mechanical case one stays in the Hilbert space, the free and interacting dynamics are given by unitary groups, and the bound states are naturally separated from the scattering states since they are orthogonal. In the case of non-linear wave equations it is not clear how to separate the initial data in $\sum_{scat}$ which correspond to soliton solutions from the initial data which are scattering states; that is part of the problem of proving $\text{Ran}\Omega_+ = \text{Ran}\Omega_-$, but $\text{Ran}\Omega \subset \sum_{scat}$. To find an example of a non-linear wave equation which illustrates these points and to develop a complete scattering theory for such an equation seems to me to be an extremely important and interesting problem.

Bibliography

[1] Bjorken, J. and S. Drell, Relativistic Quantum Fields, McGraw-Hill, New York, 1965.

[2] Browder, F., "On non-linear wave equations", Math. Zeit. 80 (1962) 249 - 264.

[3] Chadam, J., "Asymptotic for $u = m^2u + G(x,t,u,u_x,u_t)$, I,II", Ann. Scuola Norm. sup., Pisa, 26 (1972), 33 - 65, 65 - 95.

[4] Chadam, J., "On the Cauchy problem for the coupled Maxwell-Dirac equations", J. Math. Phys. 13 (1972), 597 - 604.

[5] Chadam, J., "Global solutions of the Cauchy problem for the (classical) coupled Maxwell-Dirac equations in one space dimension", J. Func. Anal. 13 (1973), 173 - 184.

[6] Chadam, J., "Asymptotic behaviour of equations arising in quantum field theory", J. Applic. Anal. 3 (1973), 377 - 402.

[7] Chadam, J. and R. Glassey, "On certain global solutions of the Cauchy problem for the (classical) coupled Klein-Gordon-Dirac equations in one and three space dimensions", Arch. Rat. Mech. Anal. 54 (1974) 223 - 237.

[8] Corindalesi, E. and F. Strocci, Relativistic Wave Mechanics, North-Holland, Amsterdam, 1963.

[9] Courant, R. and D. Hilbert, Methods of Mathematical Physics, Vol. 2, Interscience, New York, 1962, p. 695.

[10] Friedman, A., Partial Differential Equations, Holt-Rinehart and Winston, New York, 1969, p. 24.

[11] Glassey, R., "Blow up theorems for non-linear wave equations", Math. Zeit. 132 (1973), 183 - 203.

[12] Glassey, R., "On the asymptotic behaviour of non-linear wave equations", Trans. A.M.S. 182 (1973), 189 - 200.

[13] Gross, L., "The Cauchy problem for the coupled Maxwell and Dirac equations", Comm. Pure. and Appl. Math. 19 (1966), 1 - 15.

[14] Jost, R., The General Theory of Quantized Fields, Amer. Math. Soc. Providence, 1965.

[15] Jürgens, K., "Das Anfangswertproblem im Großen für eine Klasse nichtlinearer Wellengleichungen", Math. Zeit. 77 (1961), 295 - 308.

[16] Keller, J., "On solutions of non-linear wave equations", Comm. Pure Appl. Math. 10 (1957), 523 - 532.

[17] Lax, P. and R. S. Phillips, Scattering Theory, Academic Press, New York, 1967.

[18] Levine, H., "Some non-existence and instability theorems for solutions of formally parabolic equations of the form $Pu_t = -Au + \tilde{F}(u)$", Arch. Rat. Mech. Anal. 51 (1973), 371 - 386.

[19] Lions, J., "Une remarque sur les problèmes d'évolution non linéaires dans les domaines non cylindriques", Rev. Roumaine Math. Pure Appl. 9, 11 - 18.

[20] Morawetz, C., "Time decay for the non-linear Klein-Gordon equation", Proc. Roy. Soc. A 306 (1968), 291 - 296.

[21] Morawetz, C. and W. Strauss, "Decay and scattering of solutions of a nonlinear relativistic wave equation", Comm. Pure Appl. Math. 25 (1972), 1 - 31.

[22] Morawetz, C. and W. Strauss, "On a non-linear scattering operator", Comm. Pure Appl. Math. 26 (1973), 47 - 54.

[23] Morse, P. and H. Feshbach, Methods of Theoretical Physics, I, McGraw-Hill, New York, 1953, p. 622.

[24] Reed, M., "Higher order estimates and smoothness of solutions of nonlinear wave equations", Proc. Amer. Math. Soc. 51(1975),79 - 85.

[25] Reed, M., "Construction of the scattering operator for abstract nonlinear wave equations", Indiana Journal of Math. (to appear).

[26] Reed, M. and B. Simon, Methods of Modern Mathematical Physics, Vol. I: Functional Analysis, Academic Press, New York, 1972.

[27] Reed, M. and B. Simon, Methods of Modern Mathematical Physics, Vol. II: Fourier Analysis and Self-adjointness, Academic Press, New York, 1975.

[28] Reed, M. and B. Simon, Methods of Modern Mathematical Physics, Vol. III: Analysis of Operators, Academic Press, New York, (to appear 1977).

[29] Scott, C., Chu, F., and D. McLaughlin, "The soliton: a new concept in applied science", Proc. IEEE, 61 (1973), 1443 - 1483.

[30] Segal, I., "The global Cauchy problem for a relativistic scalar field with power interaction", Bull. Soc. Math. France 91 (1963), 129 - 135.

[31] Segal, I., "Non-linear semi-groups", Ann. Math. 78 (1963), 339 - 364.

[32] Segal, I., "Quantization and dispersion for nonlinear relativistic equations", Proc. Conference on Math. Theory Elem. Part., MIT Press, Cambridge, 1966, 79 - 108.

[33] Segal, I., "Dispersion for nonlinear relativistic equations, II", Ann. Sci. Ecole Norm. Sup. (4)I (1968), 459 - 497.

[34] Strauss, W., "Decay and asymptotics for $u = F(u)$", J. Func. Anal. 2 (1968), 409 - 457.

[35] Strauss, W., "Decay of solutions of hyperbolic equations with localized non-linear terms", Symp. Math. VII, Ist. Naz. Alte Rome (1971), p. 339 - 355.

[36] Strauss, W., "On weak solutions of semi-linear hyperbolic equations", Anais Acad. Brazil, Ciências 42 (1970), p. 645 - 651.

[37] Strauss, W., Energy Methods in Partial Differential Equations, Notas de Matemática, Rio de Janeiro, 1969.

[38] Strauss, W., "Nonlinear scattering theory", in Scattering Theory in Mathematical Physics, ed. J. A. Lavita and J.-P. Marchand, Reide Pub., Holland, 1974, p. 53 - 78.

[39] Strauss, W., "Analyticity of the Scattering Operator for small data", in preparation.

[40] Von Wahl, W., "Klassische Lösungen nichtlinearer Wellengleichungen im Großen", Math. Zeit. 112 (1969), 241 - 279.

[41] Von Wahl, W., "Über die klassische Lösbarkeit des Cauchy-Problems für nichtlineare Wellengleichungen bei kleinen Anfangswerten und das asymptotische Verhalten der Lösungen", Math. Zeit. 114 (1970), 281 - 299.

[42] Von Wahl, W., "L^p - decay rates for homogeneous wave equations", Math. Zeit. 120 (1971), 93 - 106.

[43] Von Wahl, W., "Decay estimates for nonlinear wave equations", J. Func. Anal. 9 (1972), 490 - 495.

[44] Heinz, E. and W. von Wahl, "Zu einem Satz von F. E. Browder über nichtlineare Wellengleichungen", Math. Zeit. 141 (1975), 33 - 45.

[45] Royden, H., Real Analysis, Macmillan, 1963.

[46] Costa, D., "Decay Estimates for Symmetric Hyperbolic Systems", Ph. D thesis, Brown University, 1972.